I Live Underwater

a memoir

I LIVE UNDERWATER

THE THRILLING ADVENTURES OF
A RECORD-BREAKING DIVER,
TREASURE HUNTER, AND
DEEP-SEA EXPLORER

MAX GENE NOHL

WISCONSIN HISTORICAL SOCIETY PRESS

Published by the Wisconsin Historical Society Press
Publishers since 1855

The Wisconsin Historical Society helps people connect to the past by collecting, preserving, and sharing stories. Founded in 1846, the Society is one of the nation's finest historical institutions. *Join the Wisconsin Historical Society:* wisconsinhistory.org/membership

Printed in the United States of America
Cover design by Andrew Brozyna

29 28 27 26 25   1 2 3 4 5

Library of Congress Cataloging-in-Publication Data

Names: Nohl, Max G. (Max Gene), 1913–1960 author.
Title: I live underwater : the thrilling adventures of a record-breaking
   diver, treasure hunter, and deep-sea explorer / Max Gene Nohl.
Description: Madison : Wisconsin Historical Society Press, [2025] |
   Includes index.
Identifiers: LCCN 2024038419 (print) | LCCN 2024038420 (ebook) | ISBN
   9781976600289 (paperback) | ISBN 9781976600296 (ebook)
Subjects: LCSH: Nohl, Max G. (Max Gene), 1913–1960. | Scuba divers—United
   States—Biography. | Scuba diving—Physiological aspects. | Scuba
   Diving—Equipment and supplies. | Underwater exploration—United States.
Classification: LCC GV838.N63 A3 2025 (print) | LCC GV838.N63 (ebook) |
   DDC 797.2/34092 [B]—dc23/eng/20241214
LC record available at https://lccn.loc.gov/2024038419
LC ebook record available at https://lccn.loc.gov/2024038420

♾ The paper used in this publication meets the minimum requirements of the American National Standard for Information Sciences—Permanence of Paper for Printed Library Materials, ANSI Z39.48-1992.

CONTENTS

Foreword   ix
Publisher's Note   xiii

Prologue   3

BOOK I: HOW IT ALL BEGAN

1 I Drown   9

2 Dick Drowns   13

3 I Live Underwater   15

4 This Is It!   17

5 MIT   26

6 Walden   31

7 My First Suit   45

8 Choppers   49

9 Pōhaku   52

10 Bubble of Steel   54

BOOK II: THE SETH PARKER EXPEDITION

11 A Dream Comes True   61

12 Disillusionment   75

13 Departure   86

BOOK III: THE *JOHN DWIGHT*

14 The *John Dwight* Murders   97

15 Man of Iron   104

16 Cape Cod Canal   113

17 Wreck Below   124

18 End of the Rainbow   134

19 The Atlantic Leaks   142

20 Cargo   151

21 The Treasure   163

22 Fo'c's'le Head   167

23 Captain Craig Arrives   175

BOOK IV: THE BENDS IN THE ROAD

24 A New Venture   183

25 Human Guinea Pigs   189

26 Gnarled on the Networks   204

27 The Elusive *Lusitania*   212

BOOK V: YOU HAVE TO HIT THE BOTTOM TO REACH THE TOP

28 Deep Water   219

29 Six Hundred Thousand Pounds   230

30 Singing for My Dinner   237

31 DESCO   242

32 I Am Theoretically Dead   252

BOOK VI: SPONGES AND CELLULOID

33 Underwater Wonderland   267

34 Sponge Diving   281

35 Tiburón   297

36 Limey Tanker   306

37 Underwater Honeymoon   320

38 550 Feet   327

BOOK VII: THE SS *TARPON*

39 The SS *Tarpon*   333

40 Pay As You Leave   343

41 The Bends   350

42 Topsy-Turvy World   357

43 The Pot of Gold   367

Epilogue: The Waters Are Still   373

Afterword   375
Index   385

Foreword

L ike many divers, I learned about the contributions of Milwaukeean and pioneer diver Max Gene Nohl much later than I should have.

I grew up in the 1970s and '80s in central Indiana, where available water for swimming was either a public pool, polluted river, or a dark and murky reservoir. While on a family vacation to Florida during high school, I begged my father to let me take scuba diving lessons only to receive his punting response of "next year, next year." In 1987, I moved north to attend the University of Wisconsin–Madison. While paging through the class catalogue, I noticed SCUBA 101, a four-credit PE elective that would result in a PADI Open Water Scuba Diver certification. For the next five semesters, I desperately tried to get a seat in the class, until I finally succeeded my junior year.

In SCUBA 101, I learned that we tread the same water "globally speaking" as the famous Cousteaus, so popular in my youth with their television show and feature films. The course had an emphasis on protecting the fragile ocean environment, and I came away incorrectly believing that most major contributions to the sport occurred in Europe or along the American coasts, far away from Wisconsin. Unless I wanted to clean up trash in our local lakes, I figured all the diving I'd ever want to do would take me to exotic warm water travel destinations.

It wasn't until a few years later, during an instructor development course, that I discovered that the diving training agency responsible for training more than 80 percent of divers worldwide—PADI, the Professional Association of Diving Instructors—had been founded in 1966 just south of the Wisconsin border in Northbrook, Illinois. This opened my eyes. And when I started looking around, I realized several of the major diving equipment manufacturers at the time were fairly local. Scubapro, a subsidiary of Johnson Wax, was based in Racine, and Dacor was located in Evanston, Illinois. My scuba cylinders were produced by Pressed Steel Tanks, and soft goods and filling equipment could be obtained from Global Manufacturing Corporation, which were both in Milwaukee. Then I learned that DESCO (the Diving Equipment & Supply Company), the longest running producer of underwater equipment, was also operating in Milwaukee. I didn't know it then, but Max Gene Nohl was one of its principal founders.

In 1993, I began training for mixed-gas diving, first through IANTD (International Association of Nitrox and Technical Divers) and then through the newly formed training agency TDI (Technical Diving International), in order to safely dive deeper than the recreational diving limit of 130 feet of water. After a few years and a few hundred dives, I was qualified to dive to 300 feet. I used my new skills to explore deep lakes and quarries, Great Lakes shipwrecks, and underwater caves of the Midwest. There were no dive shops around that encouraged or provided support for this type of exploration. In 1996, after finishing my master's degree and facing a flat job market, I opened my own dive shop, Diversions Scuba, to address this shortcoming. By 1998, when I became certified to dive with a rebreather and we began selling them in the shop, I believed we were at the cutting edge of underwater exploration—certainly a new frontier for our local diving community.

Not long after this, my eyes were opened again when I learned about Max Gene Nohl. One of my dive buddies, Chicago-area technical diver Joe Rojas, shared a grainy Xerox copy of a few chapters of Nohl's unpublished manuscript that he'd come across while conducting research at the Milwaukee Public Library.

Nohl set a world record on December 1, 1937, diving to an amazing depth of 420 feet in Lake Michigan just northeast of Milwaukee and using a rebreather suit that he invented. He used a novel helium-oxygen gas mixture and a diving timetable that was worked out by his friend and colleague Dr. Edgar End in a hyperbaric chamber at the Milwaukee General Hospital. Astoundingly, End had taken that original timetable and extrapolated to account for the exceptional depth of Nohl's record-setting dive, all while he was terribly seasick on the deck of a rocking Coast Guard cutter. And there I was, more than a generation later, taking classes in those same deep diving techniques, which were now standardized by training agencies, and using that same equipment, which was regularly marketed and sold by several different competing companies.

I have worked as a maritime archaeologist with the Wisconsin Historical Society for the last twenty-one years. As part of the Society's mission to collect Wisconsin's stories, and with the help and encouragement of Society volunteers Thomas Villand, Dr. Richard Boyd, and others, we began to accumulate stories about Wisconsin's role in early American diving history. Through this research, we found much more than those few chapters of Nohl's memoir that Joe shared with me years ago. Located deep in the Local History Collection at the Milwaukee Public Library, those chapters turned out to be just a small section of what Nohl intended to be published as a three-volume set. These are Nohl's own words telling his incredible stories.

Digging deeper into this topic, we learned so much more about the groundbreaking advances, accomplishments, and contributions that Nohl and his colleagues made to the diving industry. Most modern divers, if they know of Nohl at all, do not realize the full scope of his influence, which includes the production of rubber diving suits, masks, and fins; innovations to underwater cameras and the film industry; and significantly, more than five years before Cousteau, the invention of a "free-swimming lung" to use on an expedition in the Caribbean. Nohl's work at DESCO led to the development of underwater communications, rebreathers, and diving helmets. And together with his collaborator Dr. Edgar End, he is responsible for the rebirth of mixed-gas diving using decompression diving tables, an experiment dropped by the US Navy in 1925. These major advancements and unique designs have had a profound influence on what we still use today to explore, work, and live underwater.

While Nohl had a larger-than-life personality and was known for dramatic storytelling, one of the great mysteries surrounding him involved the book in your hands. Nohl wrote *I Live Underwater* over the course of about four years, and he was still in the process of editing the manuscript and working with an agent when his life was tragically cut short. It brings me great satisfaction to know that his stories, which put Wisconsin at the epicenter of North American diving in the early twentieth century, are finally now being shared with the world.

—TAMARA THOMSEN
MARITIME ARCHAEOLOGIST, WISCONSIN HISTORICAL SOCIETY

Nohl on board the Coast Guard cutter *Mendota* at the America's Cup races off Newport, Rhode Island, in 1934. COURTESY OF KATHY END

Publisher's Note

I Live Underwater is the posthumously published memoir of Max Gene Nohl, the diving equipment inventor, underwater film innovator, and record-setting deep-sea diver from Milwaukee, Wisconsin. Nohl was in the final stages of work on this manuscript when he and his wife, Eleanor, were killed in a car accident in 1960. After his death, the manuscript and Nohl's other papers were given to the Milwaukee Public Library Archives by Nohl's brother-in-law and co-executor of his will, J. Gordon Hecker. The manuscript remained in the archives for more than six decades. Through the efforts of Wisconsin Historical Society Press Director Kate Thompson and Milwaukee Public Library Director Joan Jacobs, an agreement was reached allowing the Press to bring this long-awaited work into print.

In the case of a posthumous publication, questions of authorial intent, publisher motivations, and editorial involvement inevitably arise. Did the author want it published? Why should it be published? And how should an editor accomplish their work in the absence of the author? Fortunately, in this case, there was no question that the author wished for the book to exist in print: Nohl had submitted his most current draft of the manuscript to a literary agency on January 18, 1960, mere weeks before his death on February 6. In addition, the book unquestionably merited publication— not just because it was a colorful, compelling, and well-written story but also because Nohl made so many important contributions to the diving world. We felt that his legacy should be preserved and made accessible, not just to those with access to the Milwaukee Public Library Archives, but to readers everywhere.

A thornier question, however, was how the unfinished nature of the text should be addressed. The manuscript had not been professionally edited, but Nohl had completed several drafts, which were preserved in his papers at the Milwaukee Public Library Archives. He had revised and rewritten the book several times, incorporating feedback from a few select individuals and refining the structure and language along the way. Yet the final order of chapters remained undetermined, and even the most polished draft was peppered with handwritten suggestions for improvement.

Though Nohl had completed an impressive amount of work—going as far as selecting photographs, writing captions, and even drawing a handful of charming sketches to accompany the text—the book was clearly still a work in progress.

Freelance editor Gary Smith transcribed Nohl's typewritten pages and organized the manuscript materials into digital files before handing them off to WHS Press Developmental Editor Elizabeth Wyckoff who edited the text and organized the art program. Kathy End—whose father, Dr. Edgar End, was one of Nohl's trusted companions and collaborators—generously provided the WHS Press with permission to reprint photographs from Nohl's scrapbook, which came into her possession after Nohl's and her father's deaths. Throughout the editing process, WHS Maritime Archaeologist Tamara Thomsen and Maritime Preservation Program volunteer Thomas Villand answered many questions and assisted in the shaping of the manuscript with their extensive knowledge of Nohl's life and maritime and diving history.

In some cases, Nohl's notes provided editorial direction and permission. In his comments on the draft sent to Franz J. Horch Associates on January 18, for example, he wrote, "Wherever you feel that the story, a chapter, or part of a chapter is weak, inconsequential, uninteresting, too short, too long, or otherwise not acceptable, I would like to have you comment on it and I would be glad to re-slant it or approach it from another angle." His acknowledgment of the text's unfinished state undergirded our decisions to make what we considered to be necessary editorial interventions. These alterations included efforts to clarify or eliminate ambiguous text, smooth transitions, avoid repetition, and correct spelling and punctuation errors.

In the interest of retaining Nohl's voice, we left some language that modern readers might find dated, such as the use of feminine pronouns to refer to vessels and the use of generic masculine pronouns. Since we were unable to pose queries to the author as we typically do, some mysteries remain, like the identities of certain characters mentioned by first name only. Also, we decided to cut some paragraphs and chapters in the service of tightening the narrative. At every turn, however, we approached this manuscript with a light editorial hand and the desire to leave intact as much of Nohl's original text as possible.

There remains one section of an early prologue that did not make it into the final text but that we would like to include in this note—not just because Nohl's statement about being "very much alive" is so poignant but also because it provides an apt introduction to the extraordinary narrative that follows.

"By the deep, six" is an old seagoing expression for "Sailor's Hell," a special Hades maintained for divers and seamen, six fathoms or thirty-six feet below the regular Hell.

Although I still am very much alive and have obviously not yet officially departed for my assigned posthumous resting place, I have spent my life searching for something in the real depths of the real sea. I could never seem to find that something. I could never even find what it is for which I have spent my life searching. All I know is that I have been through a thousand Hells in the search.

This is the story of that search. This is the story of my life underwater.

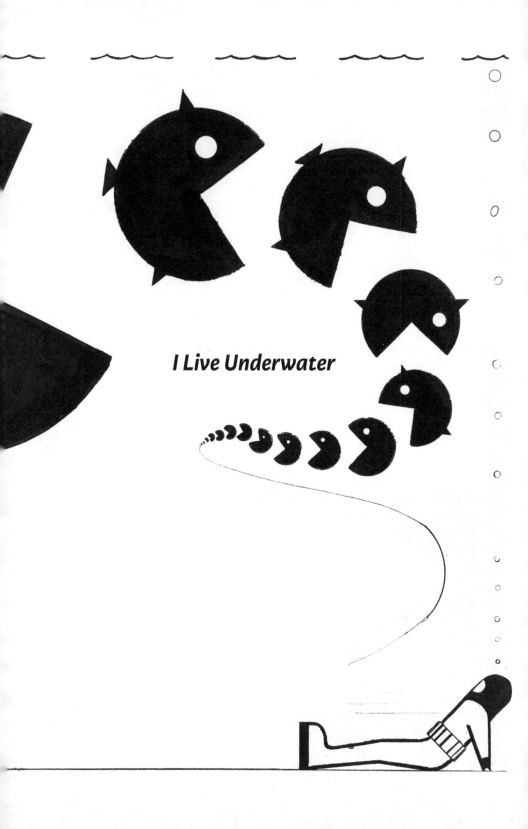

I Live Underwater

To my father

*who wanted my life to be as fine
as I know his has been*

PROLOGUE

I gasped! Suddenly, there appeared the ugliest-looking creature I had ever seen, so close I could almost touch him—a monster that could, at his slightest whim, tear me into shreds.

Tiger shark! The blood had frozen in my every vein. The wisest thing that I could do was to continue exactly what I was doing—nothing. A quick move in any direction might excite his already overaroused curiosity. As I stood there and saw that he, too, was in no hurry to make the first move, I peered searchingly through the small round window of my diving helmet to study him for any possible clue as to what he was planning to do.

He was definitely no minnow—probably a ton of fish. Looking beyond his broad nose, I could see his huge underslung mouth, his belly—glistening white like the newly painted underbody of a good-sized boat. The leading edges of his dorsal and pectoral fins loomed large in the clear water, the deceptive gray color of his entire upper body suggesting the appearance of a battleship at bay.

Flashing through my brain like the patterns of a kaleidoscope were myriads of conflicting tales as to the aggressiveness of sharks. Some prominent marine biologists had scoffed at the idea of a shark attacking a man. Others had admitted that some conditions, such as the taste of blood in the water, could excite the otherwise harmless shark into an insanely vicious attacker. Still more had mentioned undue hunger, deformities, or even personality disturbances as inciting factors. Whatever these learned scientists believed, I had on numerous occasions seen a shark strike a hundred-pound tarpon almost on the gaff with a savage crushing blow that would leave only a sheared-off head on my hook. I had glibly quoted the learned scientists on many occasions to people who had seen their companions torn into shreds.

Perhaps these incidents all occurred under unusual circumstances. But then, how was I to know whether this fellow suffered from an unhappy love life or when he had last dined?

It was true that I actually had seen hundreds of sharks in my deep-sea

ventures and had never suffered as much as a scratch from one of them, but seldom had I been on quite such intimate terms as this.

My best bet was just to wait—to outstare him as Clyde Beatty might do to his lions. And as I stared, I thought and wondered.

Just a moment ago, I had been enraptured with the beautiful world through which I was wandering—a coral reef offering as peaceful and breathtaking a sight as mortal man may ever behold. My eyes had been startled by the infinite shades of dazzling yellows, purples, oranges, and greens of the weirdly shaped formations through which I had been climbing. Myriads of brilliantly hued fish had gaped at my Gargantua-like form with curiosity. Then, as inquisitiveness turned into alarm, they had with a snap of the tail darted away to more certain pastimes.

Now the tables were turned—my 185 pounds were facing perhaps a ton of formidable fish. I was a man—one of the slowest and clumsiest of all animals, inadequate in my own domain, but here, in a foreign world in a bulky two-hundred-pound diving suit, almost completely helpless. Man has always been forced to rely on his comparatively highly developed brain to gain superiority over his environment. Here, I had only my ridiculously inferior strength and speed against that monster before me.

I wondered, as I had seldom done before, just *what* was I doing down here? It was true that I had good reasons for this dive—searching for commercial wool sponge beds—but there must have been countless other, better ways for me to make a living. For a quarter of a century, I had followed this strange profession—twenty-five years of suffering, heartbreak, and misery. And yet, I knew that I had found, down here in this crazy world, that something for which I had always looked—that something for which many people search their entire lives and never do find.

The shark incident, to this story, is inherently meaningless. He looked at me—I looked at him. Finally, he tired of staring and simply swam away. I turned and went about my business. But after that, I thought a lot. Why, really, was I down here in this fantastic world?

The answer to that question is the story of this book—the story of my own lifelong search for adventure.

Book I

How It All Began

1

I Drown

It was one of those hot, sultry days in July when just breathing the burning air was torture. Scattered groups of middleweight overaged vacationers sat and soaked in wet cotton clothes behind constantly wagging fans in the shade of the many massive trees in the hotel yard.

If it was torture, we hadn't noticed it particularly—we had been too busy hunting since breakfast. Returning early for lunch, we suddenly became conscious that it was hot from the groans of the wilted group, and all of us welcomed the idea of a cool swim before the noonday meal.

This was a July of many years ago. I was about ten years old at the time—so it must have been around 1921—and, as usual, was spending a portion of the summer at the little hotel on Moose Lake, one of Wisconsin's thousands of beautiful freshwater glacial lakes.

I watched the rest of the vacationers glibly dive from the pier end into the cool, refreshing water and, as each head bobbed up, invariably, a cry would come down at the sheer joy and exhilaration of the cold plunge.

My entry was somewhat different. Not having as yet learned the art of swimming, I found it necessary to confine my immersion to the wading process.

Moose Lake is an unusually deep lake, apparently spring fed, and its water is considerably cooler than that of most of its neighbors. Accordingly, the swimmers soon found themselves amply refreshed and returned to the hotel located well over the top of the high bluff and out of sight of the pier.

I soon found myself alone as I slowly progressed through the process of gradually wetting my hot body the hard way, the icy water level climbing painfully upward as I walked out from the sloping beach. However, I was so hot that I was determined to wade out to neck level and stay there just a little longer than I could possibly stand it. By so suffering, my blood and body would get sufficiently overcooled that the heat would be welcome for a few hours.

I had been warned a thousand times—Moose Lake was treacherous! As the banks rose steeply up from the water's edge, hemming the lake in like a deep-set jewel, so did the lake floor fall steeply off into tremendously deep water. The ledge on which I stood formed the so-called bathing beach but was actually only a deceptive shelf.

At the end of the pier, one could easily look down into the clear spring-fed water on a calm day and see the submerged drop-off disappear into a mysterious azure blue nothing. "The lake is bottomless," 'twas said by the local sages.

I knew this all very well, of course. How silly it was to even think about it. I was not going to keep on walking away from the shore once the water had reached up to my neck. I wasn't crazy!

Soon the water level had reached my armpits. I was so very tall that I suffered a lot more torture than most boys my age would have to reach the corresponding levels on their bodies.

I'd take one more step out—I could wade in up to my neck. Up to that time, I knew no particular fear of nor any particular fascination with water.

To my mind, already thinking in scientific terms, it was merely H$_2$O, and in this case, a very pleasant heat-absorbing medium.

However, that next step placed my foot over a turn in the bank on a slope of more grade and several inches deeper than I had expected. At that degree of submersion, the water had so much buoyancy that I could not maintain sufficient weight upon my feet to gain the necessary traction to regain my equilibrium.

I found my nose underwater then took another desperate step—forward but down the falling slope—in a futile attempt to arrest my fall. A gasp for air with my head submerged—water in my lungs—violent choking and gagging—further falling forward—water in my lungs with loss of my natural body buoyancy—and then falling—falling—falling—down the submerged slope into those blue-black depths. Then a vivid kaleidoscope of events raced across my mind.

I never was very certain about the popular phenomenon that a drowning person may enjoy his last few minutes of consciousness watching a mental motion picture of his past life. But there it was!

The pictures that raced through my mind at lightning speed did seem to refer to the more colorful parts of my past life, but seemingly only one phase of it was strongly stressed—the manifold warnings that had been instilled into my mind about the treachery of that lake. Everyone at the hotel had repeated an individual form of this warning over and over again.

These had all been of passive importance to me, it seemed, but nevertheless they must have made deep grooves in my young mind. Now, to find myself falling helplessly down that slope—I couldn't seem to grasp anything, however desperately I tried, except for that elusive and crumbly sand and loose gravel.

My fear seemed to be swelling to colossal proportions—the clutch for anything solid—the frightful realization that I had been completely alone and no person had seen me disappear—the vivid mental picture of that huge deep abyss—the countless warnings against this very thing—the torture of my violent, uncontrollable gasping of water—and then, as my conscious mind appeared to be dimming, so another previously hidden mind seemed to be coming into its own, thinking with an almost transcendental clarity and power—a frantic, delirious fear of death. And then, as it reached that infinity that is beyond infinity that is still beyond that infinity ad infinitum, there seemed to be a change.

I was gagging violently. Groups of poorly defined forms were huddling over me. In what world was I now? I thought I could hear my father's voice calling my name. The light was so bright in my eyes.

None of them knew how long it had been. They had been pumping for a long time. They were never too confident that artificial respiration would do. But now, I was showing signs of consciousness. I was coughing voluntarily and helping to rid the water from my lungs. Everything would be all right.

Someone had walked down the dock to smoke a cigarette. A casual reflection: "Wasn't that Nohl boy still in wading when we finished our swim?" A quick check, and I was not in my room. One of them was an expert underwater swimmer.

A half hour later, it was all over except for a headache and a sore feeling in my chest. But in my mind was a mark that would never be gone. This, both literally and figuratively, was the start of a career in deep-sea diving.

2

Dick Drowns

As the days rolled on, I found that the incident did not become merely a passing memory. A thousand times over, when my mind was free in a world of dreams, I'd find myself falling again down that slope into the mysterious blue-black depths of water. My heart would pound as I cautiously ventured out to the end of the pier and, on clear days, looked down into the abyss below.

It seemed as though a deep groove had been cut into my boyhood brain. It might not have made much real difference in my later life. If I'd become a lawyer, like my father, I might never have had occasion to associate with any body of water larger than an apartment bathtub—and even there, if necessary, I could have a shower. But another incident only a few weeks later was to add a vivid intensification to my drowning experience.

Dick Moody was one of my closest friends. Vacationing simultaneously at Moose Lake, we had planned our entire lives together in our intimate boyhood conversations. Through our common thoughts, ambitions, likes, and dislikes, we had built up a warm companionship that had meant a whole world to each of us.

Dad unexpectedly decided to split our vacation at Moose Lake that summer, returning to his office under the pressure of business. I said goodbye to Dick for a few weeks only to spend the time in Milwaukee planning some of the things that we would do when I could return. These weeks seemed without end, but finally my family and I were on our way back to the lake.

Arriving in the hotel yard, hardly before the car had come to a stop, out I jumped looking for my pal. "Where's Dick?" was my first question, as I failed to see him on the grounds.

A solemn expression fell over the faces of the group that had assembled to greet us. After a pause, one of them volunteered, "Dick hasn't been found—he was last seen in swimming the day before yesterday."

The blood froze in my veins—"in swimming"—that slope!

Dick's body came up the next day.

I saw him there, floating face up. He had come up underneath a cluster of rowboats, seemingly trying to squeeze in between two of them. A cross-lake wind had kicked up a ruffle of water. The buoyant boats were briskly bobbing up and down, and the dark waves rocked Dick's body with as impassionate a regard as if he had been one of the water-soaked logs that frequently mingled with the boats at the crowded landing. His pale white, deeply pocked body was badly bloated.

All of this, superimposed upon the vivid memory of my own experience in exactly the same place just a few weeks earlier, caused deep water to take on a new significance.

I was afraid.

3

I Live Underwater

A few more years slipped by. I had learned to swim, under professional guidance at Bechstein's Swimming School in Milwaukee. More summers had been spent at Moose and other lakes, and shallow water was gradually relinquishing some of its menace. But never could I look down into the blue-green water without feeling a pang of terror strike through my body.

At Moose Lake, a rowboat had sunk. She wasn't much of a boat—apparently so dilapidated that no one thought her worth the toils of salvage. She had been a little bit rotten all over, and into portions of her soggiest parts, concrete had been poured to cover up the leaks and hold the hull together. Still, she was leaking badly and weighed down with concrete and waterlogged wood. It was no wonder that she had gone down.

The resting place that she had selected was conveniently located in shallow water on the ledge—perhaps not over six feet deep—so that I could easily swim down to her and practice holding my breath by hanging on to a gunwale.

One day, during such a practice dive, I found that when I reached my hand up into the inside of the boat, my forearm was in the air! The boat had settled upside down and across an uneven bottom in such a way that I could swim under her port side and up into the interior.

I could breathe!

Here, a fathom under the surface, I could live. Perhaps a couple of cubic feet of air had been trapped under the inverted hull, but this was enough sustenance for several minutes.

Completely accidentally, I had found the essentials of a diving bell or caisson. However, as I had no means of replenishing the air, the first few

minutes that I had enjoyed in the underseas were about the extent of the air's usefulness. But in those few minutes, in which I could sustain myself in that world of which I was so afraid, I found that I could play with stones and shells on the bottom and yet continue breathing with my head in the air.

Although this was totally impractical and nothing further developed from it at the time, here was a violent stimulation to my imagination and an experience that was one of the emotional highlights of my life.

And so, with this incident, the first chapters of my life were written and then completely forgotten for half a dozen years, except for an occasional strange feeling that crept over me at the sight of deep blue-green water—a feeling of stark fear.

4

This Is It!

J ack Browne, too, liked anything that was unconventional, unusual, or unexplainable. His numerous exploitations in this direction had, from time to time, come to my attention—he also lived in Milwaukee and was just a few years my junior—but fate had not as yet arranged to cross our peculiar paths.

Until now.

Jack had run across an article in some popular science magazine on "How to Build a Diving Helmet."

He was not a firm believer in following superior instruction to the T, even with a diving helmet, when life and death were at stake. In fact, the resemblance to the carefully worked out plans in the article was practically negligible. However, there it was, almost completed, and just bristling with the Browne brand of engineering.

The essential structure of the helmet was an old, discarded five-gallon paint can. Fitted to its front was a piece of glass that he had found down in the basement in one of his father's storm windows— but then, it was summertime. A garden hose (it had been raining quite a bit and the lawn seemed to be getting along quite well by itself) would carry the air, a tire pump found in the garage would serve as a compressor, and some of the window weights from his house (it was summer and the windows should be open anyway) would supply the ballast.

Of anyone I have ever known—and that includes

one of the world's rarest collections of crazy people—Jack had to bow to few for top honors. Particularly contributing to this was a psychological phenomenon—the kid seemed to be utterly devoid of the sensation of fear. I don't believe this basic instinct ever as much as entered his mind.

He had the helmet just about as far as it was going to progress in construction. He hadn't gone down into the water, not in any degree from a sense of fear but more probably due to a lack of understanding as to just how the thing worked.

This seems to be something which is universally a little difficult to understand in the popular mind—what keeps the water out of this open-bottom type of helmet? Perhaps a short dissertation on the principle of this simple but basic diving device would be pertinent.

A drinking glass inserted upside down into a dishpan of water will trap the contents of air in the upward portion of the glass. This captured air cannot conceivably escape, as its pressure holds down the water and keeps it from entering the dry interior of the glass.

As this same glass is inserted into deeper water, the increasing water pressure will tend to compress the trapped air upward into the glass; at thirty-four feet there will be only half a glass of air, the water level rising into the lower half.

A simple demonstration of the elements of a diving helmet may be carried out with rather dramatic results as follows: A lighted candle is mounted on a flat cork or small block of wood in such a way that it will float in a basin of water. An inverted glass is lowered over the candle and pushed as far as desired down into the water, with the small candle-bearing raft continuing to float on the surface as determined by the imprisoned bubble of air. Far under the water, the flame continues to burn!

This rather impressive sight will last just a few seconds, however. The flame will be seen to flicker more and more feebly as the oxygen is consumed in the small interior of the glass until it expires. But, by connecting a rubber hose to the glass and supplying fresh air at a pressure enough in excess of the water pressure for it to flow, it would be possible to keep the candle burning as long as it lasted.

The analogy to the simple open diving helmet is almost perfect. The outfit consists of an air compressor or pump, an air hose, an inverted vessel large enough to accommodate a man's head, and lead or iron weights to correspond to the downward force exerted by the hand in pushing the glass down. Lastly, the diver himself is the equivalent of the candle, as the metabolic processes of his body burn and turn the food he has eaten into heat and energy.

Jack's helmet, theoretically, combined all of the necessary elements. From a standpoint of good engineering, however, it was a mess. The window weights were merely tied with string to the frail metal paint can. The window was bolted in the front in such a way that any prospective diver would have to keep in mind and be prepared for its falling completely off at any moment. The air hose connections were soldered on with such a delicate bit that it would be necessary to use extreme care in handling it—one bump and the hose would fall off.

But when I saw that helmet, crude as it was, my heart jumped into my throat. I had read about such devices, knew that they existed, and understood how they worked, but I had never actually seen one. Here before my eyes it was—brand new, even though made of old junk. As a diving apparatus, it was still to see the water for the first time.

The thought of descending in it was almost too much. The dread of deep water, racing madly through my brain many years after the Moose Lake experiences, came vividly back.

What would it be like to actually go down into that mysterious world? To be able to breathe, walk, see, use my hands and feet?

The answer was clear—I was scared stiff! But here was a challenge. Jack was willing to let me take the first dive.

My raft was anchored about half a mile off the shore of Fox Point in front of my family's summer home, about ten miles north of Milwaukee on Lake Michigan. Conditions were not ideal, to state it conservatively— a brisk east wind whipping across the lake was kicking up quite a chop. The big square raft, which I had built out of eight fifty-five-gallon oil drums mounted under a heavy four-by-four bolted frame, was not exactly designed for comfort in a heavy sea. As we approached her, she was bucking like a bronco, her windward side triumphantly climbing to the highest crest of each oncoming comber as her leeward side lunged into the preceding trough, followed a second later by a complete vice versa.

Boarding her was no minor feat, raft and approaching boat each being a serious menace to each other. She was now like a wild horse feeling its

The helmet used on Nohl's first dive. Left to right: Max Nohl, Jack Browne, and Verne Netzow. Nohl wrote, "We later attempted to make this into a complete suit, patching together a pair of old overshoes and the rubber from several discarded inner tubes. Comparatively, a sieve is watertight." *MILWAUKEE JOURNAL* PHOTO first rider, grimly determined to dump anything that wasn't bolted to her frame into the sea as she rolled through a fast arc of almost ninety degrees.

Wearing an old pair of tennis shoes and swimming trunks, I ventured to get into diving position, gingerly mounting the frail diving ladder, which had been intended to enable swimmers to board the raft. This brought back some of the sensation of old cattle ranch days, riding a steer with only a rope around his belly to hang on to. That ladder was plunging up and down like the hammer of a giant punch press, alternatingly dunking me violently into the lake well over my head and then snatching my wet body completely from the water up into the direct blast of the mercilessly cold offshore wind, down into the drink and up again, over and over, as the boys on deck, unable to stand, crawled on their bellies in an attempt to get the seventy-five-pound diving helmet in position to drop on my shoulders.

This they finally succeeded in doing with no infrequent bursts of unprintable words that would put any sailors to shame as the loosely hung window weights swung and mashed their fingers.

I think that Jack had dangled eight of those window weights from the lower rim of the inverted paint can helmet. I think eight because I counted the bumps on my head later. As they lowered the trap over my head as carefully as they could, first the four front weights would swing back and klunk me on the nose and forehead as hard as I tried to steer my nose between the second and third ones—and then the raft would pitch the other way and again—*klunk!*—the four back weights would chime in from behind on my unpadded skull. And thus it continued as they—so painfully cautiously—lowered the hat inch by inch.

My troubles were quickly terminated, however. Someone slipped, the seventy-five-pound apparatus came crashing down on my shoulders as the ladder end of the raft rose to meet an oncoming wave, and the bottom of Hell gave way.

My 185 pounds, plus the hammerlike impact of that 75-pound helmet, plus the tremendous increase in gravity as my side of the raft leaped to the huge breaker beneath was just too much for that broken-down little ladder. The raft rose and most of the ladder rose, but the most important part to me at the moment did not rise—the rung I was standing on.

Thus, all of my elaborate plans for scientifically getting into the water had gone awry. I was well started on my forthcoming career—well on my way down to the Deep Six.

I had intended, as I had read all divers do, to grasp the descending line before releasing my hold on the ladder, so that I could use this rope like a trolley to control my rate of descent in case I had difficulty in adjusting myself to the increasing pressure. This was very important. I had carefully thought out exactly how I would transfer my weight from ladder to rope once my helmet was submerged.

But now, I was just falling. None of my carefully laid plans seemed to have the slightest application. I groped frantically in three-dimensional space for that line but couldn't find anything nor see anything. And then— as I was trying to make up my mind whether I was alive or dead or in trouble, and if so, exactly what I could do about it—something seemed to come up and hit me.

I was, for the first time, on the bottom!

I couldn't see anything—just a blue-green color. Then, I realized that the window was badly fogged and that I had read in a book what to do about it. The water level, forced down by the compressed air, was over my chin and so little below my mouth that it didn't take much of a nod forward to enjoy a drink of water if I should so choose. That was the remotest thought in my mind, but I found that by sucking in a large mouthful of Lake Michigan and then puckering up my lips and putting a little compressed lung air behind it, I could form a nozzle, which made a first-class job of washing the entire window of its translucent but nontransparent fog.

As I spewed that stream of water, as if by magic, the fog melted into nothing—and there before my eyes was the most amazing sight that I had ever seen.

It was beautiful—breathtakingly beautiful! But it was more than this to my bewildered eyes—here was the world for which I had always sought.

Men spend their entire lives looking for a few simple, basic things—a lifework, a philosophy, a true love, a few but not many other fundamentals. Not many ever find these things—perhaps even fewer have any idea for what they are looking—but they waste their lives in search.

I, too, had spent a restless life searching for I knew not what. And then, suddenly, here I was, face to face with that intangible something for which I had been looking. It didn't take months or days or even minutes to decide.

The instant that fog cleared and allowed me to peer into that beautiful blue-green world, I knew *that* was what I wanted. From the standpoint of the submarine sights that I have seen in subsequent years, this was probably a rather dull comparative spectacle—the sandy floor of a not-too-clear freshwater lake. A few rather droll, drably colored fish curiously gazed into the window of my helmet. The scraggly freshwater flora, which are able to grow on the sand and shell bottom, waved gently to and fro with the water particle action.

But here was another world! A world as apart from earthly experience as if I had been allowed a similar visitation of another planet.

Here were all of the elements of the romance of man's scientific accomplishment. I recalled that Jack's paint can–window glass–garden hose–window weight apparatus, based on a commonly known principle, was not exactly an engineering novelty nor masterpiece, nor would its completion be recorded in the annals of scientific progress. But it worked—here by a combination, however crude, of man-made materials and the application of science, I was able to invade another frontier. The sight that I was viewing very probably never before had been beheld by the eyes of a human being. The world that I was in was part of a great virgin territory—the almost totally unexplored ocean floor, which covers approximately three-quarters of the surface of our earth, about three times as large as all the dry land put together, a world that humankind knows almost nothing about to this very day.

Flashes ran through my mind of the uncounted billions of dollars in lost shipping lying down here, of fabulous treasures that had disappeared beneath the waves, of almost fantastic but unquestionably known great sunken cities, of even possibly lost continents, of inconceivably vast natural resources. Had I heard it said, "Nothing left on this earth to explore"?

I was truly in another world. I was trying to discover for myself how to learn to live again. I found great difficulty in getting traction on the bottom. I seemed to have no weight. My sense of gravity was entirely different. I would reach for something that I could see. I couldn't touch it. My sense of vision had to be relearned. All of my other senses, by which mortal man has his only contact with the material world, were operating in a vastly different and confusing way. I would have to relearn to live. This was a very exciting thought.

Perhaps more important than anything else, here I was in this world of water of which I had always been so afraid. Fear was turning to fascination.

I was breathing. I was comfortable. I was perfectly safe (if that window didn't fall out, I would be safe, anyway) in a well-made helmet.

Though not yet twenty, I had lived a life in which I had sought many thrills. The degree of the thrill had always seemed to be the degree of the danger. The thrills of racing automobiles, low-flying airplanes, speedboats, or just the roller coaster at State Fair Park had, in the final analysis, been nothing more than in proportion to the sense of danger they invoked. Whether real or merely sensed—whether in defiance of basic fears of speed, heights, or physical injury, or whether in defiance of code or strong arm of the law—this joy of adventure was always related to the danger involved.

Here, in this helmet, I was in a world of which I was morbidly afraid. Here was a normal instinctive fear of water, which every human experiences to some degree, exaggerated into a deeply embedded phobia.

Here was the answer to my adventure-crazed mind.

I had not noticed that my teeth were chattering almost violently enough to shake out a few fillings. My body was vibrating like a Model T fender. Lake Michigan had not yet warmed up for the summer. I had better go back before I shook right out of my skin.

After a short search, I found the descending line—on which I had failed to descend—and made two unsuccessful grasps for it with my air-conditioned sense of where things should be in space before actually touching it on the third.

Here was my first reminder of the world I had left above—the line was jerking up and down violently with the pitch of the raft overhead. I grasped it quickly and tightly and was yanked off my feet as she jumped to an oncoming sea for what I thought was the start of my journey back. That portion of the journey was very short-lived, however. As hastily as the line had whisked me off the bottom, I was deposited again on the bottom—mine and Lake Michigan's—and I now had a few more souvenirs to add to the bumps on my head, only on that above-mentioned portion of my anatomy that I couldn't display quite as proudly to my friends that evening.

A moment's reflection suggested that by catching the line as high as I could reach at the lowest portion of its cycle, I could, by intermediate climbing, be on my way. And then, a few moments later, looming up before the helmet window was the bottom rung of the ladder, which I grasped as I released my grip on the rope. Climbing up this friendly looking but rickety structure, I soon found my helmet breaking the surface and my eyes squinting from the bright surface light. I felt the boys grasping the helmet top and, as they braced themselves, lifting its seventy-five-pound weight off my head. Already frozen, now coming up wet into that icy blast again, I probably was about as miserable as I had ever been, but inside of me, a fire was burning furiously—I had found what I wanted.

5

MIT

College was in the air. All of my friends were talking about it. The advantages of one over the other were weighed in all-knowing authority a thousand times over. There was a lot of talk about football teams, fraternities, college rules, and coeducation.

My father had very generously and open-mindedly given me the privilege of choosing any college or university in the country and of selecting the course of study I wanted. In high school, I had found most of my studies not too hard, but also of very little interest. I had had a few brilliant streaks in physics, algebra, geometry, and science subjects, but the fact that these corresponded to my nonacademic interests was purely coincidental. I had often wondered if I would not be much happier as a plain bum than as a bum with a college degree.

But, now, it seemed so vastly different. I had an ambition—a terrifically big ambition—and now the only decision to be made would be: where could I get the finest training to help me reach my goal?

I had learned from my history book of the great Industrial Revolution— how, in the past few generations, the application of scientific research had revolutionized the entire life of man. The pace of progress had been set at a high gait in almost every field of endeavor, but somehow deep-sea diving seemed to have been left far behind.

Way back in 1837, an Englishman named Augustus Siebe had invented the first diving suit. A century later, there was little evidence of improvement. Siebe's original helmet was almost indistinguishable from the latest models I had seen pictured in books. It is true that a few accessories had been added—some modern suits carried telephones and some were fed air by engine-driven compressors rather than hand pumps—but beyond that, there was little to show for an intervening century of progress.

Had Siebe been such a genius that he had developed the ultimate at the very outset and thus made impossible improvement over perfection? From the staggering figures depicting the economic challenge of the great ocean floor to man and from the unbelievably brief history of his few and feeble attempts at conquest of this world, the answer seemed to be obvious.

The problems, it seemed to me, were not problems to be solved by greater brawn, enlarged lung capacity, or more resistant bodies—they were problems of hydraulics, physics, chemistry, and mathematics. In short, problems of engineering. The exploration of my generation would be vastly different from that of our rugged forebears. No longer would man's penetration of the Arctic depend on his ability to withstand the terrific rigors of the polar climate. No longer would tropical expeditions depend upon man's sheer endurance against merciless heat and disease. His progress in the few remaining undeveloped portions of our earth's land surface would depend upon equipment—science and engineering.

There were hundreds of colleges and universities where I could learn engineering. Many of these undoubtedly offered very fine courses. But there was one name that would invariably come up on top—one school that would always immediately be mentioned in the answer to my frequent question, "What is the finest engineering school in the world?"

"The Massachusetts Institute of Technology!"

I learned that their entrance requirements were very stiff, that they had the reputation of offering the toughest courses of any college in the world, and that only a very small percentage of those who succeeded in entering (the cream of the cream of the crop) ever made the grade.

It would mean College Board Examinations in everything, and so I dug in. The exams came, and I passed them—at least most of them. I failed on my required foreign language, but then MIT agreed to let me in if I took a summer course in it.

MIT meant Boston (the institute is actually in Cambridge, on the opposite bank of the Charles River), and Boston to me meant, first of all, the Atlantic Seaboard. I had never seen the ocean, and soon I found, the first day in the Hub in the fall of 1929, that it was not very easy to see. I found great grim freighters from many lands snuggled into slips highly walled with buildings, scummy canals, dock-bedecked rivers, everywhere the smell of fish, but no sea. I couldn't seem to get sight of a wave as big as I had seen on Moose Lake, the city being settled back in a deep bay well-dotted with protective islands. However, a few days later, I found that for two dollars I could go deep-sea fishing. In the company of a score of other landlubbers, I soon found a deck heaving under my feet as our craft answered to the incoming ground swell, salt spray in my hair as we plowed into a brisk chip, and no trace of land appearing to the east as we chugged beyond the outer fringe islands.

Boston soon came to mean several more things to me. There was the great technical library at the institute and the fine Boston Public Library. I eagerly scoured every page of everything in the index relating to "deep-sea diving."

And then downtown, in the heart of old Boston, was Andrew J. Morse & Son, Inc., the largest of the two manufacturers of diving suits in the United States. Mr. Lawton, the manager, was very friendly and invited me to see the first diving suit on which I had ever laid my eyes.

It was beautiful! It was the most beautiful thing I had ever seen. I must have stood there with my mouth wide open as I stared at that highly polished, chromium-plated, heavily begadgeted helmet—this monster was the key to my future world.

But it would be rather an expensive key. That outfit—with the necessary hand pump, hose, shoes, lead weights, and accessories—listed at $1,325.00.

Thirteen hundred and twenty-five dollars! Where would I ever get that much money? I tried a little mental arithmetic—college graduates, I had been given to understand, going out into industry in those days should not expect much more than twenty-five dollars a week for their embryonic knowledge, until such a time as they have acquired an accumulation of "practical experience." If I could live on twenty dollars and could consistently tuck five dollars each week into the bank, it would take me exactly five years to place the order for that suit, which, with a college course intervening, would be nine years from now. That was not too stimulating a thought. My deep-sea ambitions did not look exactly sunny.

Perhaps I could buy a used one. However, browsing along the New England waterfront, I soon learned that diving suit manufacturers did not follow the plan of the automobile companies by putting out a new model each year. The last model had come out in 1837—there had been none before nor after. They did not wear out, particularly, being made of copper, brass, and lead (except, of course, the rubber dress, a relatively inexpensive portion of the complete outfit), and the used ones were worth almost as much as the new ones. The one other possibility, that of making one, was even more remote—the dies, patterns, molds, and intricate machine work would cost more than Morse's price.

Then I met Vose Greenough. In appearance, he was a typical Harvard student of that time—hair clipped about half an inch long, making his head resemble a hairbrush; dirty white buckskin shoes (for year-round wear and never, under any conditions, cleaned); brown pants; and a gray coat. It seemed that most Harvard students that I saw followed this same pattern of dress.

Vose was studying biology—that was quite evident upon entering or even approaching his room in the dormitory from the assortment of dried dead cats, weird-looking skeletal remains of almost anything that could die, and smelly jars laden with assorted internal organs. They didn't seem to me to be particularly attractive decorations for a bedroom and study, but apparently, they were to him. However, of concern to me was, one, the fact that he was also studying with particular interest in marine biology, and, two, that he, the son of a prominent and well-to-do Boston (Brookline) family, owned a complete diving suit, which he had purchased to go down and make firsthand observations of his finny subjects.

I told Vose about my dreams and problems. I don't think he had previously heard much enthusiasm on his own method of making direct observations of his marine biological specimens in their own habitat. And so, we struck a common chord. Vose liked diving, too, more than just as a means to an end—he was much interested in the engineering aspects of the art. In him was a deep streak of adventure—he loved nature and he loved the sea. Diving was his hobby.

My highest hopes, in looking up the biological student at Harvard who owns a complete professional diving suit, had been to just see the equipment and to talk with him about diving. But now he asked, "Would you like to take a dive in it?"

"Would I like to take a dive in it? *Would* I like to take a dive in it? Hell yes!"

He asked about the time I had available, being familiar with the long hours of the MIT schedule.

"Anytime, day or night, classes or no classes, whatever is going on, I'll make time," I answered enthusiastically.

"How about tomorrow?" he blurted.

6

WALDEN

There are always some men who find happiness not in the companionship of fellow humans but in the solace of living a hermit-like existence apart from the rest of the world—men who find spiritual contentment in the lavish beauties of Nature, in the masterpieces of God.

Such a man was Henry Thoreau. In 1845, he moved himself bodily and spiritually from the civilized world of his day to the innermost chambers of God's domain on earth—Walden! His memoirs of those years, his book *Walden*, will forever be loved as a classic of early American literature.

Walden, today, a century later, could still be the spiritual retreat of any soul who might seek an escape from the turmoil of our modern civilization. Only thirty miles from the great teeming center of metropolitan Boston, there immediately exists for the casual visitor the same atmosphere, as foreign to the adjacent bustling city life as another heavenly planet might seem to one stepping from a rocket ship.

The commonwealth of Massachusetts, aware of the rare natural beauty of the location and its historical effect on American literature, has taken it under its wing for everlasting protection.

Walden Pond, in the terminology of New England (though a pond has always meant to me not much more than a good-sized puddle), is the unpretentious name for this beautiful, deep blue-green lake. Conspicuously lacking any trace of the usual fringe of ramshackle cottages, dance pavilions, and Coca-Cola signs, it still nestles languidly, like a great sapphire, within the virgin sylvan beauty of the surrounding forest.

The thrill that most people must experience when first viewing the breathtaking beauty of Walden was somewhat diluted for me by several more practical observations. I had just discovered that the entire lake was frozen over solidly. Chopping through the ice, I found that it was about ten inches thick, plus a layer of several inches of snow over that.

Looking up from the hole I had fashioned, I noticed that I was standing immediately adjacent to the outer end of the only man-made structure visible on the lake, a wooden pier. Nailed to its end was a sign: "25 Feet Deep." Another glance indicated that the pier was about twenty-five feet long. It didn't take much advanced trigonometry to recognize that the submerged slope fell off at an angle of about forty-five degrees.

I described, as indifferently as I could, my mathematical calculations to Vose, but he gleamed only more—and he had been gleaming ever since we left Boston—at the pleasant prospect of again working with the diving equipment. He explained that was why he had selected Walden as an ideal testing location—it was possible to get to almost any depth desired. The bank sloped down precipitously into hundreds of feet of water. In fact, he had always heard that the lake was bottomless—at least, no person had ever been successful in touching portions of the lake floor with a sounding lead. To find equivalent depths in the ocean would require an extensive offshore voyage, plus the expense and inconvenience of chartering and working from a boat. Here, even in the summer without the convenient ice covering on which to walk, it was possible to find deep water by walking out from the shore with no need whatsoever for a ship.

I thought that that was "fine" and that he certainly had discovered the "ideal spot."

It might have been beautiful Walden to millions of Thoreau fans, but to me, it was beginning to look like my memories of Moose Lake.

The temperature was only a few degrees above zero. The merciless wind, sweeping furiously across the trackless snow cover of the lake, seemed to blow through every stitch of clothing we wore. Every movement that I made was with reluctance, as the penetrating wind took advantage of any gap at neck, wrist, or waist of my clothing. I was shaking with cold, and the thought of going down into that ice water was far beyond the limits of comprehension of my benumbed brain. I just went ahead unloading the heavy diving gear from the car and lugging it out to our location on the ice, as though I really were going through with it. I would have to take off every stitch of clothing I was wearing and then crawl into that icy rubber dress, which was now almost as stiff as a Victrola record at this temperature.

I perhaps was so cold and miserable that I didn't have the strength to suggest that we do it some other time or under some improved conditions. I felt like a man preferring to die by freezing rather than make the effort to stir even if it could save his life. It was easier just to go ahead as planned. Undressing, now down to nothing but a pair of colored glasses, the thought came to me that I was standing on a public pier, but modesty didn't seem to be worth a turn of my head—not even a winter bird would stray out today. The first clothes to go on were two suits of iced woolen underwear. Following this came the diving dress, which was now almost brittle enough to crack.

Vose had finished chopping a more substantial hole through the ice where I had made the first test of its thickness. It was about two feet in diameter, which seemed to be just large enough to pass the widest portion of the diving suit.

The diving dress into which I was wiggling resembled somewhat a baby's sleeper suit with built-in feet. It was made of a double layer of light canvas, separated by an impregnated film of rubber, making the finished material a little more than one-sixteenth of an inch in thickness. Normally, it is soft and flexible, as it would be soon from my body heat. This dress forms a waterproof envelope for the diver's body, allowing him full freedom of movement of arms and legs but in no way protecting him from the water's pressure.

This always seems to be a source of misunderstanding, even to the extent that many people worry about the horrible results of a tear in the dress, visualizing the pressure rushing in and crushing the diver's body. Actually, this portion of the diving suit merely transmits the external pressure to him, and its presence offers nothing more than a separating film between the supposedly warm and dry interior and the cold and wet exterior. It protects the occupant no more from the weights of that water than a silk stocking would protect a person's otherwise bare foot when accidentally stepped upon by a careless heavy boot.

The only parts of the anatomy to protrude from the airtight envelope of dress and helmet combination are the hands. In most diving work, it is considerably easier to work with them bare than through a thick rubber glove or mitten that might otherwise be an integral part of the dress. A watertight seal is established at the wrist by a tight rubber cuff, which must be lubricated with a smear of soap suds for the entry of the hand. This was the next process—pouring soapy ice water on my wrists. In a moment, I vaguely felt what seemed to be my numb hands slip through the elastic cuffs with an audible snap.

The only other opening in the diving dress is the place of entry, which is bordered by a thick gum rubber collar. This serves as a gasket in connecting the dress to the helmet.

Now that I had properly squirmed into position, the breastplate, or lower portion of the helmet, was dropped down over my head, the cold brass touching my neck at various points with no particularly pleasant effect. The twelve holes punched into the rubber collar were then slipped over the corresponding studs bristling from the lower rim of the breast-plate. Heavy brass flanges were placed over these same studs followed by wing nuts, which snugly drew flange to breastplate with the rubber collar gasket between them. This made a watertight sandwich between rubber and metal. I learned then—the hard way—when tightening these nuts, always turn the heavy metal wrench away from the diver's head rather than toward him so that when the tool slips it doesn't make a dent in his skull.

Accessories followed next. The lead shoes, weighing about fifteen pounds apiece, were strapped to my feet. These were immediately useful to establish a little more stable ballast against the vigorous wind. Water-proof mittens, used by divers only in extremely cold water, followed—this, if ever, was certainly the place to use them.

The helmet, with front window removed so that I could breathe fresh air until the last possible moment, followed. This screwed to the corre-sponding portion of the breastplate and sealed with an eighth turn very much like the breech of a large gun. Now, for the first time, I was protected from the wind. Except for the small open window, the suit was an airtight and windtight seal, and I started to feel the first signs of life returning to my numb torso.

I couldn't help but admire Vose's painstaking caution and thorough checking of every piece of the equipment, which gave me considerable confidence for this first experience. He had gone so far as to drag a boat out to the edge of our hold and place the air pump in it—just in case, by some strange freak of nature, that heavy ice should break up, the pump would be safely floating at the surface and not sliding down a careening floe to join me at the bottom for what might be quite an extended dive. Even though that ten-inch thickness *seemed* strong enough to support an armored tank division, I developed an even higher respect for that man Vose.

He now asked me to sit down at the edge of the hole, dangling my lower legs in the water. Inserting my already almost-frozen feet in ice water required a forced effort, but it was followed by the very gratifying realization that I was not sensing any additional discomfort. The water, although as cold as water can get—thirty-two degrees—and to the imagination about the coldest thing in the world, was actually thirty degrees warmer than the air. The thermometer was hovering around the temperature of two degrees.

Vose then called for help on the belt, and a moment later, I thought a truck had run over me. I had already been carrying the weight on my shoulders of the breastplate and helmet, which totaled about fifty pounds. But that had been heaven compared with the gravity-loving thing they were slinging around my waist. This was called, I had read, a belt, but at the moment that didn't seem to be a very appropriate name for it. The lead weights, in ten-pound units, totaled one hundred pounds. They were actually supported not by the belt to which they were bolted but by suspender-like straps that crossed over the shoulders of the breastplate.

I wish I had realized at the time, but here probably is the terror of deep-sea diving at its peak. Vose had just screwed the front window or faceplate in place, and I was completely sealed in my new rubber-, copper-, and glass-lined world. My terror probably was the result of a number of factors.

The pump was shooting a deafening blast of air into the helmet interior. This roar, plus the terrific crashes as the boys clanked metal to metal in checking over the equipment, was magnified to titanic proportions by the high degree of reverberation within the small sealed space, enough in itself to scare a fellow half to death. Then (although I was still in air, except for my lower legs), as the faceplate was closed, there was an immediate indescribable sensation of oppression from the tiny increase of pressure within the suit. This was small compared to what the deep water was soon to offer but decidedly noticeable to some strange subconscious sense as are minute changes in atmospheric barometric pressure. In this case, it was only the air building up a slight back pressure against the stiff folds of the dress, but it was sufficient enough to emphasize the instinctive fear of enclosure, or claustrophobia, which is probably innate to some degree in everyone. To further this fear was the blindness resulting from the heavy fogging of the windows. My warm moist breath and body humidity instantly froze upon contact with the windows, which like everything else within sight—with the exception of the water I was about to enter—was thirty degrees below freezing. Thus, my vision was limited to the dimly lit protuberances mounted in the helmet interior. I was so cold that I wondered if I would be able to stay conscious, a thought which did not lessen my anxiety about the forthcoming dive.

Each of these things individually, or even collectively under different circumstances, would have aroused no such degree of emotion in me, but now, added to my old fear of that black water beneath me, they were stirring new things in me that I never knew were stirrable. The terrific weight of the suit, over two hundred pounds, seemed to be irresistibly dragging me downward into that terrifying water. Only the slippery ice seat, on which I wasn't certain I could remain from one second to the next, was separating the two worlds. Exhausted by the heavy work of lugging the equipment, exhausted by the strain of carrying the brunt of the suit weight, and exhausted by the misery of the cold and my fear, I felt, as I had never felt before, a feeling of utter futility—that weight dragging, dragging down into that black hole, almost as if a giant winch were pulling me downward toward a slow torturous death. I was almost beyond the breaking point in my emotions—I didn't care if I lived or died. Death would be welcome to escape this agony.

But I was determined to go through with it. Probably primarily motivated by the less-than-noble virtue of desperation, I slid off my ice seat and found myself falling. I couldn't see where or how, but in a split second, I saw the water level passing over my window and knew that I was submerged. With half a turn, I blindly felt the descending line and grabbed it, and then as Vose felt me shift my weight to it, he slacked all my lines, and I was on my own. Hand over hand, I could now lower myself to the bottom at any rate that I chose.

Nohl diving under
the ice on a later
visit to Walden Pond,
December 1933.
COURTESY OF KATHY END

I dropped down several feet to avoid bumping my helmet on the edges of the ice hole and then stopped for a second or two to get organized.

This was now a complete diving suit, and I could suck up no water to spew on the windows, so I just licked the drops of condensation off with my tongue. Streak by streak, the panorama of the underwater world opened up. There really wasn't much to see, but it was breathtakingly beautiful to me. The bright yellow descending line loomed up against the black water like the hawser for the *Queen Mary*, and I wondered how I could get my hand around such a huge rope. A glance downward with a little squirm revealed that my hand was equally magnified, looking more like a baseball catcher's mitt than my little paw. The pain was terrific. Overconfident, I had been sliding downward into pressure a little too rapidly and had not been able to check myself in time. I felt as if someone were driving a chisel into my head with a sledgehammer.

I remembered all of the instructions as to what to do. Vose had pushed a piece of gum into my mouth immediately before the faceplate had been inserted, and I started chewing on this violently. Nothing happened, and I chewed even more violently. Still, nothing happened.

Then I started blowing my nose, finding considerable difficulty in the process with my hand on the outside of the helmet and my nose obviously on the inside. However, I found that I had a little set of muscles that I never knew existed. By puckering up my mouth and flexing these muscles at the same time, I could largely seal off my nasal passages, and then by blowing could simulate a fine "blow."

It didn't work at first, but on the third blow, suddenly—click!—the pain vanished completely, and I was ready to be on my way again.

Down and down and down, into that inky black water, I slid for what seemed to be an endless distance. But just as I started to wonder if Vose had accidentally picked out that "bottomless" part of the lake, up came a steep, soft, oozy mud slope to meet my feet. It was really a slope, too; I soon delightedly found that my trigonometric calculations were, in practice, correct—the angle was just about forty-five degrees.

It was then that I made my big mistake. I let go of the descending line. Why I did this, I'll never know. I had read in my many perusals of the diving books that this was very bad practice, but somehow I had too many things to think of, and the feeling of contact with the bottom gave me a sense of confidence and suggested that I could just stand still for a moment while I organized my thoughts as to what to do next. Anyhow, I let go of that descending line, my magic thread back to that tiny hole leading to and from this ice-covered world.

The bottom, which I had just so warmly welcomed as the objective of my dive, actually had little to offer as a bottom. Sloping precipitously down into the unplumbed depths of Walden, the floor's oozy surface was covered with a treacherous carpet of slimy leaves, which offered as much traction as an equally steep mountain of ice. The air pump had not quite been able to deliver air as fast as the increasing pressure had required it, and my diving dress was pressing into my body with almost a crushing force. In addition to having difficulty breathing, I had very little buoyancy to offset the lead weights, and I was extremely heavy on my feet.

I wasn't on my feet for long, however. They were out from under me as though I had stepped on a banana peel, and down that slimy slope I went. I tried desperately to grab something but found little purchase in frantically clutched handfuls of leaves and ooze.

This, if ever I have experienced it, was a feeling of utter helplessness, almost like falling out of an airplane. I didn't know how to operate the valves, how to handle myself in the water, how to compensate for the change in vision, how to do anything except instinctively, like a drowning man, grasp for any straw within reach. The pressure was compressing my chest so heavily that it had pushed most of the air out of it, and my efforts to inhale were like trying to lift the weight of a ten-ton truck with my diaphragm. If only I had that line, I could have stopped myself with that. But now there was nothing. Frantically, I groped. Down I went into the black abyss.

My hand finally found a firm, rough object to which I could cling, which later proved to be the branch of a sunken log. To this, I held for dear life and waited. I was desperately short of air. Physiologically, I should have been panting, but because of the pressure on my chest, I couldn't even inhale, much less pant. This was as miserable a conflict as I had ever run into to date.

However, as I waited, the pump apparently was catching up with the increasing back pressure of my suit, and slowly, because I was not going deeper, I felt the air filling up the upper dress and again allowing me to breathe, although the air was very foul. I now had a free arm to reach for the air hose and give the signal—three long pulls, "Haul me up."

A moment later came the most gratifying feeling I could ever have imagined. The answer came down from the faraway world above, three long slow pulls, which meant, "We will haul you up."

Immediately after that, I felt the lifeline drawing taut, and soon I found myself being towed up the slope down which I had just slid.

I now had my first chance to look upward as they towed me. This was the high spot of the day. The ice covering the lake was blanketed by several inches of snow, making it as opaque as if it had been light-proofed for a photographic darkroom. But in brilliant contrast to that great jet-black sky over this world of water was a bright circle of light, blazing like the moon through the thin, clear mountain air of a Colorado night. This was the hole through which I had come. Appearing no larger than a dime at this distance through the crystal-clear quiet waters of Walden, it was like the end of a tunnel—a tiny, gradually enlarging spot of light penetrating a stygian gloom. And, as the imaginative locomotive engineer might occasionally wonder, I, too, questioned how I could possibly hit that tiny hole.

A long pencil of light thrust itself down into the inky waters and shed a faint and mysterious trace of illumination onto the steep slope. Almost vertically under the hole, I now saw the descending line ahead and, in a moment, was able to regrasp it with a conviction that I would never again make the mistake of losing it—that magic thread leading back to my known world.

A tiny leak had developed at my collar, probably as a result of the strain on the dress from the chest squeeze at the bottom of my fall. It had reached the proportions of a trickle. I had previously been quite comfortable, having warmed up from my shakes enough so that I even felt sorry for the fellow who had to stay up in that penetrating wind and who couldn't enjoy the restorative warmth of the ice water. Yet, the trickle was so paralyzingly cold that, with little imagination, I could envision it as a red-hot poker branding a mark down my back. A little pool had collected in my right foot, and again, physics or no physics, I was ready to admit that there is nothing any colder than ice water. (Actually, it is the conduction of heat from the body rather than its low temperature that makes the ice water seem so cold. In a dry diving suit, no part of the body is in direct contact with the water.)

I found that my suit was getting so light as the dress expanded with the decreasing pressure that it was necessary for me to valve air. I was anxious not to "blow up"—I most probably would miss the hole we had cut in the ice and bump my head on some other underpart of the surface, which at this time of year would be as soft as concrete.

Through the upper window, I saw the tiny circle of light gradually enlarge as I approached it, and in a few minutes, I was to it, through it, and out of the water. I thought that I had been miserable with cold getting into the suit, but I soon found that that had been a summer picnic in comparison. I was now wet inside with ice water and chilled to the very marrow of my bones.

Vose told me the length of rope he had paid out. I calculated, again summoning my mental trigonometry, the depth which I had reached.

I had been in 120 feet of water.

Walden Pond was too darn cold, but summer was too darn far away. I had heard that down in the basement of the giant MIT Engine Laboratory was a virtual river. This was used for cooling the huge engines. Since I was registered in mechanical engineering and would soon be working down there, I decided to investigate.

I found that there was a small man-made river down there running under heavy steel gratings. There was one place where it would be possible for me to lift out a grating and, by a tight squeeze between a couple of massive power plants, get down into the water with just about a foot drop from the floor to the water surface. I found a piece of rope, tied to the end of it an old connecting rod, and thus improvised a sounding lead. The water in this particular spot, a somewhat deepened well or reservoir, was twenty-five feet deep.

Although I knew that the water would be almost as cold as that at Walden, at least I would have a warm place to dress and undress. Permission was granted, and we moved over the entire gear.

I had been able to squeeze through the descending location to take the sounding, but I soon found out that I hadn't made allowances for the bulky diving helmet. It just would not go through. After all of the laborious dressing operations, the boys now had to take off the helmet and breastplate and then squeeze me through the location and then redress me. We were enjoying all of the luxury and spaciousness of two people doing the same thing in a typical telephone booth.

However, after much grunting and straining, I was ready for my number two dive in a professional diving suit. I had since thought over very carefully the numerous mistakes I had made on the first dive and resolved to have this one go smoothly.

As I slid off the concrete floor, I made sure that I had the descending line in my hand as advised in the books that I had read. The little hand pump was supplying a loud, pulsating, reassuring stream of air, and all seemed well.

The pump was small, and I had no excess air, the twenty-five-foot bottom being about the limit of the portable apparatus. Thus, I had little buoyancy and started sinking like a rock.

Almost instantly, I was in total darkness. The light was none too bright in any part of this basement. What there was seemed to be shielded by the big engines behind which I was diving. Worst of all, the water was inky black and would probably shield out even bright sunlight.

This was my first experience in total darkness. There was something terrifying enough about the tiny confines of the diving helmet, which seemed to be accentuated by the small space in which I was working, but now to find myself totally blind was a sensation on which I had not counted. I frankly was scared stiff. Probably only because I dreaded even more to admit it and go back and face the jeering skeptics who were waiting for me up topsides did I keep going. With the minimum of air, I was heavy, and only by a real effort could I stop my rapid sinking.

The bottom never seemed to come. Had I made a mistake on the sounding? It seemed, as I was falling down a bottomless black shaft, that it could only mean certain death.

Then suddenly, a reassuring soft mud came up to meet my feet. I immediately felt better. My fears vanished.

I spent an hour down there, occasionally forcibly blinking my eyes in order to tell whether they were open or closed. I experimented with every possible adjustment on the diving suit, trying various combinations of open and closed positions on the control valve and on the exhaust valve. I maneuvered around the confined space on my knees, belly, seat, and back. I turned upside down, purposely letting the air run up into my legs, a rather dangerous situation in a diving suit, and then practiced the trick, which I had read about, of trying to get them down. This consists of attempting, through a violent straining effort, to get into a knee chest position so that the air will, by stages, spill out from the feet into the knees, seat, and then back up into the helmet portion of the suit. Since there is no way to exhaust air from the feet, it is possible for a diver to get so much air in his legs that it could be impossible for him to right himself. If there were leakage water in the suit, it would be particularly embarrassing since it would run down into the lowered helmet and, if in excess of a gallon or two, drowning would result.

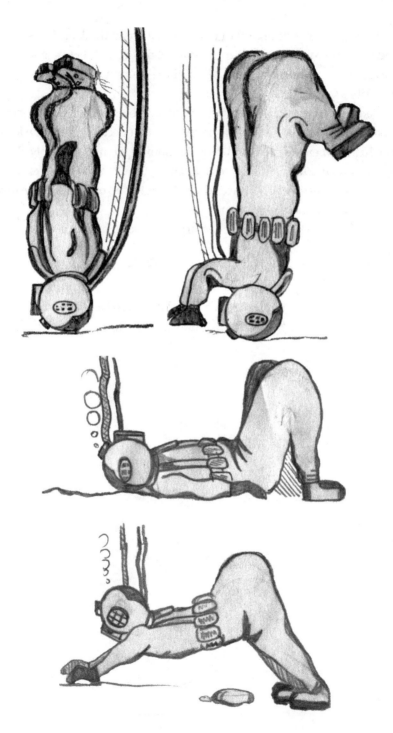

I ran through a rehearsal of everything I could think of. I was satisfied that I now was complete master of the diving suit. Before going up, I decided to make a brief exploration of the bottom of the well, just for curiosity's sake.

Suddenly, down in the soft bottom, I felt a big, metal heavy object. I picked it up and started the fascinating game of deciding what it was by feel. In a moment, I had it—it was a big monkey wrench, probably about eighteen inches long. I tied it to my belt and then, a few minutes later, decided that I had accomplished what I set out to do on this practice dive and gave the "Haul me up" signal.

Topsides, we had to reverse the process, taking off the helmet completely before I could get out of the tiny space.

I displayed—and saw for myself for the first time—the prize. It wasn't much of a wrench, the handle being bent and the entire adjusting mechanism being badly rusted. No one wanted it, and so I assumed title to the tool. I cleaned it up and put it in good working order and enjoyed years of use from it.

It was my first loot from Davy Jones's Locker.

7

MY FIRST SUIT

One of Vose's friends, also a student at Harvard, had caught the fever and had also bought a complete diving suit. He, too, was a biology student and was eager to observe his subsurface subjects in their natural habitat.

Steeped in theory, armed with first-class equipment, and possessing high degrees of intelligence and good judgment, they carefully checked every piece of their equipment and thoroughly learned the signals. To any lay observer, they were well prepared for their first dive.

Choosing a location off the coast of Massachusetts, which they knew was particularly teeming with marine life, they loaded their gear into a small boat and proceeded to the diving location. Anchors were laid in the approved seaman-like manner. The day was perfect. The sky was clear. There seemed to be no possibility of a sudden squall or any other exigency that would introduce an element of potential danger.

The proud owner slipped into the dress, and in customary order, his tenders carefully and evenly tightened the breastplate nuts, preparing him for the water with a fastidiousness that would make a hardened professional diver chuckle. All hands knelt low as he moved toward the stern for final adjustments preparatory to going overboard. Finally, there was the "start pumping" command, the window went into place, two knocks sounded on the helmet top for the "ready" signal, and down he went.

The sounding lead had just indicated thirty-six feet—it was "By the deep, six!"

They had a fine little pump, a two-cylinder brake job built solely for portable diving operations. According to the instruction sheet, this apparatus is not recommended for depths in excess of thirty feet. However, it is common practice to use this size a little beyond its limit, since the next larger pumps get into terrific weights and are not practical to transport.

This was his first dive. A little excited, naturally, he was probably burning considerably more oxygen in his overstimulated body than the typically lethargic-minded professional diver. Also, upon reaching the bottom, he probably did not have his dress properly adjusted and was suffering from the energy-sapping condition of excessive buoyancy. In this state, lacking traction on the bottom, every move is always a strenuous and futile effort.

This can be nothing more than surmised, but these two factors plus the undersized pump must have resulted in a substantial shortage of air and resulted in labored breathing.

Of all experiences that seem to bring about fear to a novice in a diving suit, probably nothing exceeds that of sensing an air shortage or actual air failure. This, above everything, seems to symbolize danger—being weighted to the bottom of a watery grave with hundreds of pounds of lead, with suddenly the dreaded specter of suffocation lurking insidiously ahead. Thus, a vicious circle is immediately established: a slight shortage of air (which normally could be offset by the trained diver with a few moments of complete relaxation) causing increased breathing, the increased breathing causing apprehension, the apprehension or sense of fear increasing the diver's metabolic rate as the body's protective machinery shoots adrenaline into the bloodstream, and the increased metabolism immediately demanding a proportionally larger quantity of air. Thus, air shortage causes fear, and fear increases the demand for air with a repetition in the same circle, often up to disastrous proportions. Sudden fear, it has been shown in laboratory experiments, can skyrocket one's oxygen requirements as much as 1000 percent, which, with an already insufficient supply, can result in a serious situation.

The conclusion, aside from this particular experience, is this: When you are working on too small a compressor or for any reason have insufficient air, assume a state of suspended animation. Try to act mentally like a hibernating bear making a belly full of food last for the entire winter.

If the air compressor breaks down or for any other reason the air is stopped, just say to yourself: "I'm out of air and I don't care—I'm out of air and I don't care." And mean it! And if it gets so serious that there doesn't seem to be a chance of escape, just say to yourself: "I'm about to die—so what!"

Thus, Nature's system of supplying a man in danger with tremendous adrenaline-stimulated energy reserves for the combat of self-preservation, although one of the miracles of design, is not very desirable in a diving suit. In fact, it is so dangerous that an excitable man should never, under any possible adverse conditions, go down in a diving suit.

Anyway, our novice diver found himself panting violently. He gave his three-pull signal—"Haul me up"—but was in such a hurry to get started that he even further wasted his short air reserve by attempting himself to climb the descending line. At the surface, there was considerable confusion as to how it would be best to get him back into the boat.

However, he had no time to wait while they on deck were going about the process of being confused. He reached up, extending his hands out of the water to grab the gunwale of the boat in an apparent attempt to lift himself sufficiently high so that they could open his faceplate.

A diving suit rapidly assumes a tremendous amount of weight as it comes out of the water and loses buoyancy. Thus, this effort to get the faceplate, and accordingly almost the entire helmet, above water level, required almost an upward lunge.

Things then happened fast. The small boat listed sharply toward the suddenly applied weight. The tenders, standing, fell and added their weight to the depressed starboard side. The air pump, scaling probably two hundred and fifty pounds, lurched downward. The boat sipped a huge gulp of water.

Further fell the boys. Hundreds of pounds of water were pouring in. And then the worst happened—the madly careening boat spilled the air pump into the sea. Down it went like a plummet, entangling the diver in the snarling coils of hose.

This must have been a horrible end—already desperate for want of air and then suddenly and irresistibly dragged to the bottom by the heavy pump, pinned down without any possible chance of help for a torturous death by suffocation.

Sometime later, his body and the equipment were retrieved with grapnels.

That was my first diving suit.

I bought it cheap from his parents—for twenty-five dollars, to be exact—including the curse. I was now a fully equipped, professional deep-sea diver.

8

CHOPPERS

"**A**re you the deep-sea diver?" queried an excited voice. One of my fellow freshmen had answered the telephone and had called me at our fraternity dinner table, reappearing with a very puzzled look on his face.

"Yes," I now answered into the phone, outwardly confident but inwardly wondering just exactly what the dozen or so experimental practice dives in Walden Pond really did make me.

"What do you charge?" demanded the voice. I mentally reviewed the vague references to standard rates of pay for divers I had seen in the library books.

"That depends on the nature of the job and the time involved," I replied, trying to be as unspecific as I could until I recalled just what divers did charge.

He went on to explain that he had recently been fishing off Marblehead (a few miles north of Boston) but, to a large extent, had really been merely sitting there thinking and relaxing. He had a fishing pole in his hand as the usual excuse for his popular pastime of doing nothing. Suddenly, the unexpected happened—the thing that he dared not hope for—he caught a fish!

In the sudden excitement of the moment, he lost control of a number of his ordinarily well-regulated muscles. Out of his gaping mouth and overboard went his false teeth.

He had raked, dragged, scooped, sworn, and prayed, but all that he had retrieved was a pathetic collection of small slime-covered rocks, shells, two Coca-Cola bottles, and a rusty tin can.

The choppers had set him back $250, every penny of which he had paid in cash. However, this was a small part of the real value. The suffering that he had lived through to break those store snappers in he would never again endure for an additional $500. Thus, they had a value to him of over $750.

The word had leaked around town that there was some fellow at MIT who was a diver and, stretched a little from mouth to mouth by the time the story reached him, that this delver into the depths was a first-class professional.

This was no time to insist nobly that the banner of pure truth be waved high, and so I stated, noncommittally as to my standing in the profession, that I would do the job on a minimum rate charge, that is to say fifty dollars, for which I would either find the teeth or spend sixty minutes underwater looking for them. This was perfectly agreeable to him, since he knew within a few feet where he lost them and figured there was little chance I wouldn't see them in an hour's time.

I had no trouble rounding up a few helpers—there must have been something fascinating about watching a diver go down, for they were always ready and willing to go along. I instructed them to look and act as bored as they possibly could on the job—not to ask any questions and to keep conversation at a minimum—to make it look as though we had done this many times before. They behaved beautifully.

My plan, a reproduction of what I had read in Tom Eadie's exciting book *I Like Diving*, was to tie a hunk of iron to the end of the descending line and drop this at a spot that my client pointed out as the most probable location of the teeth. To this weight, I tied another rope, which was the circling line. Using the weight as the center of the area to be searched, I made a circle at arm's length from it, and then fathomed out another six feet to make a second circle in the opposite direction (to avoid tangling descending line and air hose). This I continued, treading out a pattern of concentric circles to cover the bottom in a thorough search pattern of decreasing probability, six feet being sufficient increment in radius to allow me to see or feel and overlap each added area.

The water, stirred up by the backwash of the breakings rolling up on the beach, was sufficiently clouded so that it was necessary for me to traverse my geometric area on hands and knees with my helmet window about two feet above the bottom.

Then, suddenly, a sight appeared that scared three years' growth out of me. As I lumbered along like a big black bear in a coal bin at night, there, just in front of me, loomed up tusks like a streak of lightning in a midnight sky. Magnified tremendously by the water, only a few inches from my window, and glistening brilliant white against the deep somber-colored bottom of rotten leaves and mud, they veritably gleamed, as inside did I, also, at the successful completion of my first job.

As I, a few minutes later traded the teeth for my check, I grinned again. I was now a professional diver!

9

Pōhaku

"Three hundred feet beneath your own two feet is a treasure of twenty million dollars. It's yours, if you can get it."

Pōhaku was listening with eyes like saucers as the captain augmented his startling statement with further detail: "We are just about passing over the charted location of the Cunard liner *Lusitania*. She has been lying in her grave below us since 1915. That's a day that the world will always remember. Two German torpedoes tore open her starboard side and, in their concussion, they shook the entire world."

Pōhaku was on a small ship skirting the Irish coast on his way to England to stroke the American Henley Regatta. There were over three hundred feet of Atlantic under them, but deep water held no terror for him.

Ernest Johnstone was his real name, but such a prosaic designation utterly failed to describe this colorful fellow. He had been born and lived his entire life in the Hawaiian Islands but, through the whims of fate, was now a fellow student as well as a fraternity brother of mine at MIT. Quite in contrast to the typical non-athletically inclined tech engineer, this super-man had a build that would put a gorilla to shame. Probably resulting from a youth spent largely in the water and on the surfboards at Waikiki Beach, he presented a physique in my observation comparable only to actor Johnny Weissmuller playing the part of Tarzan.

Pōhaku in native Hawaiian means "rock," and it was a highly suitable name for him, both in the intended sense of the hardness of his giant muscles and the sheer massiveness of his torso. It was appropriate, also, in his highly developed ability to swim underwater. The expression "he can swim like a rock" actually was a tribute in his case.

Pōhaku could disappear underwater, and long after the time we were fearfully certain that he was dead, up would come his grinning head. After a quick breath, he would submerge again for an equally long sojourn. After observing his athletic skills and aptitudes at hockey, swimming, crew, and boxing (whereas most MIT students find it difficult to make time for even one sport, he took them all up simultaneously and was usually the outstanding man in each), it was not difficult to believe his stories of fighting sharks underwater equipped with only knife and goggles.

Thus, the realization that only three hundred feet of water separated him from the twenty-million-dollar treasure of the *Lusitania* was a fascinating thought to Pōhaku, who could do almost anything better than anyone else. The world's record for deep-sea diving was 306 feet, which, strangely enough, had been made practically in his front yard in the warm waters off our own famous Pearl Harbor. It was back in 1915 that the United States Navy diver Frank Crilley had reached this depth and established the record, which was still unchallenged.

I thoroughly liked Pōhaku. Both of us being nonconformists, we decided to get an apartment together on the Fenway, breaking away to some extent from the conservative traditions of our fraternity and yet living close enough to it that we could keep our fingers in all of its activities.

In our isolated abode, we talked much of the sea and particularly of the great unexplored world beneath it. I had serious hopes that this natural fish-man would go far with me toward the fulfillment of my dreams, which seemed to have a tremendous appeal to him. However, fate (or more specifically, the dean at MIT) was unkind—Pōhaku flunked out and went home on the sore-eye special. MIT agreed to grant readmission to him a year later and Pōhaku came back, but then he flunked out again. It wasn't that he didn't possess the mental capacity—his studies were just interfering with those five sports, the bull sessions, the parties, and the girls.

Pōhaku went back to his beloved Hawaiian Islands. I was not to see him again for a score of years, but he had left something behind—he had fanned a fire that was now burning furiously within me. The *Lusitania*!

Just six feet beyond the world's deep-sea diving record was lying that giant hull, undisturbed in her dimly lit water grave, carrying the secrets of the first great world war and a treasure estimated to be in excess of twenty million dollars.

10

Bubble of Steel

The *Lusitania* was lying at a depth of 312 feet.

The world's deep-sea diving record, as we noted, was 306 feet, a mark that had been set in 1915—by strange coincidence, the same year the *Lusitania* had sunk. The difference was only six feet—one little fathom—but it wasn't this single fathom that made the difference. It was the practicability of doing useful work over an extensive time under those conditions.

Perhaps a finer diving suit should be invented. But, actually, in spite of previous comments about the lack of improvement over the period of approximately one hundred years since Augustus Siebe had invented the modern and yet first diving suit, there didn't seem to be much that could be improved. The fundamental problem was the pressure, and the diving suit made no pretense of protecting the diver from it—it merely offered him a dry envelope in which he could breathe and which was sufficiently flexible to allow him to use his arms and legs. Perhaps improvements could be made in offering better window designs, easier dressing, or greater resistance to abrasion or tearing of the dress, but fundamentally the modern diving suit was a pretty darn good piece of apparatus.

There had been another line of endeavor on which my now voluminous files were veritably bulging—the steel diving suit, a pressure-withstanding apparatus subjecting its occupant to no more than normal atmospheric conditions. There were scores of patents on variations of this method, and I had found many accounts of the performance of the few that had actually been built.

The lightest and cheapest of these that I had come upon tipped the scales at fifteen hundred pounds and sold for around thirty-five thousand dollars. If they accomplished their purpose, this investment would be a small tribute toward such staggering treasures as were waiting to be salvaged.

However, they did not. To allow movement of the arms and legs, ball-and-socket joints were provided, usually at the hips, shoulders, knees, and elbows, depending upon the amount of freedom the respective inventor wished to provide his delver of the depths. These joints operated very much like the eyelid would over the eye, and they must have been machinist's nightmares to manufacture—thus, the cost.

At two fathoms, the better of these suits could accomplish its function—the terrific weight was lifted by the buoyancy, and the diver could, although in an extremely cumbersome fashion, perform possibly useful movements. It was reported that one diver had actually succeeded in tying a knot in a rope with his pincer hands.

However, two fathoms is no problem for the rubber-suited diver under the water pressure, nor even is twenty fathoms (120 feet) for a good deep-water man. But the steel suit, descending down to and beyond this average practical limit for the rubber apparatus, is completely helpless. The tremendous external pressure squeezes those ball-and-socket joints so tightly together that a man with the biceps of a blacksmith could not move his arm. Thus, the diver is frozen into a figure with the flexibility of a statue. Design that joint whatever way you will, but you will still have high-pressure water outside pushing against low atmospheric pressure inside.

The only other way to go down was in a bell, like Beebe's Bathysphere, and withstand terrific external water pressures. But then you'd give up any possibility of the two important functions—maneuverability and dexterity, or in other words, the use of hands and feet. Beebe's Bathysphere was lowered on a cable from a rolling, pitching, drifting surface vessel and offered no control of position other than "up" or "down," and the signals were telephoned up by its occupants. They could see out of a window, but it seemed that the important thing was not to just see something, but to be able to accomplish useful work—cutting, dynamiting, and working with scores of underwater tools.

Perhaps the answer was better divers. But comparing my physique with that of Frank Crilley, it was quickly obvious to me that I was not going to be the guy. It was apparent that I was not going to have many choices in a search of the world for mightier men than Crilley. And he had lost consciousness, I understood, after two minutes at 306 feet on his historic record dive.

If any, there were going to be two answers to the problem.

The rubber suit possibility—to find out why it is, from a physiological standpoint, that a man cannot stand these terrific pressures and to determine whether it would be possible to alleviate the effects of these conditions, realizing that to have full natural flexibility, the diver must bear the full pressure of the water on his body.

The pressure-withstanding equipment possibility—to find out whether, by housing the diver in a pressure-withstanding vessel, some satisfactory means could not be found to provide him with maneuverability and dexterity.

The latter was my first approach and was the solution I sought in my baby bell.

There are three kinds of buoyancy, which I should like to define, since these are expressions I will frequently have occasion to use. *Positive buoyancy* is the distinguishing term that we might give to anything that tends to float, such as a cork pushed under water. *Negative buoyancy* is the opposite and would apply to anything that sinks, such as a rock dropped in the water. *Zero buoyancy* is the intermediate status—the object referred to doesn't want to sink, it just wants to stay where it is, like a fish. Grasp that, and you can call yourself a submarine engineer.

As I plodded through my courses, extracurricular activities, fun, and the various development of my little professional diving business, racking my brain was that one burning question: How can I get dexterity and maneuverability without subjecting the diver's body to pressure? It was insolvable—until, all of a sudden, one day right in the middle of calculus class—it was clear as a *bell*!

I would build a hollow steel sphere—the most perfect shape to withstand pressure—and design it so that it would have zero buoyancy, not negative buoyancy like Beebe's Bathysphere, but as an inert body in the water. It wouldn't sink. It wouldn't float. It would just stay wherever it was. A steel-coated bubble of air suspended motionless in three-dimensional space.

There would be electrically driven propellers—three reversible screws to provide motion in any conceivable direction. These propellers would be driven by externally housed motors so that stuffing-box leaks would not admit seawater into the interior of the bell.

The control stick part was what I liked best of all about it. It was designed so that whatever direction the diver moved his hand in three-dimensional space, rheostat-controlled electric motors would instantly move the bell in the same direction.

The bell was designed primarily for observational and directional purposes, such as the placing of dynamite charges, grapnels, clamshell buckets, etc., since the diver could place himself wherever he wished and, over the telephone, could direct the handling of heavy equipment from surface vessels. A steel arm with magnetically operated jaws would also be fitted to the front for the occasional opportunity that the occupant might have to pick up some small object.

The bell itself was to be made of manganese steel of such that she, because of her perfect spherical shape, could theoretically reach a depth of eight thousand feet. Dr. Beebe's Bathysphere, probably the only other practical diving bell that had ever been built, had plunged to a depth of 3,028 feet—over half of a mile—and made the only invasion at these great depths that the world has ever known. In New York, the following year, I met Dr. Beebe and found him as cordial as his deep-sea exploration had been revolutionary, and also Otis Barton, his co-diver and engineer of the Bathysphere.

The whole plan of my proposed baby bell, free to move at will in any direction in space, sounded almost out of this world, but as fantastic as it sounded, careful check and analysis of every principle involved over and over again by myself as well as by a number of hard practical-minded engineers could not reveal a fallacy, and so ahead went her construction.

Before the summer ended, she was finished and diving in the waters of Lake Michigan. But soon, fall approached and it was again time to go back

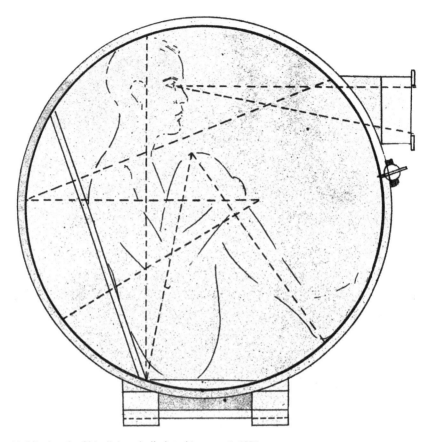

Nohl's sketch of his diving shell, dated January 5, 1935.

to MIT—the bell, the *Lusitania*, and efforts in that direction would have to wait for the next summer. And so, it was back to Boston, the rubber suits, weekend diving jobs or just experimental descents if business was bad, and, of most immediate importance, the biggest task that I had now—to get my diploma.

Book II

The Seth Parker Expedition

11

A Dream Comes True

A few weeks after the bell had been laid up for the winter on the shores of Lake Michigan, and we were in full swing back at the MIT grind, I picked up a current issue of the *New Yorker*. Casually paging through it, my fancy was captured by a chance article—the story of an almost fantastic expedition.

The radio star Phillips H. Lord, whom I had heard many times on the air as the nationally loved character Seth Parker, was going to fulfill a boyhood dream. He was planning to sail around the world in an old windjammer with no purpose other than the pursuit of adventure.

I was not surprised to note, among the other colorful plans, that he intended to search for some of the old Spanish galleons with their fabulous fortunes—pirate gold.

My eggs were cold; my mouth was hanging open. The whole world had come to a stop as I read about this—a real expedition that was going to sail in the next few weeks.

What would it be like to actually sail on that ship? The thought was so fantastic that I hardly dared think about it. I had a nine o'clock class—I had almost two more years of college ahead of me before the diploma was mine. How would a fellow go about getting started to apply for a job on an expedition like that? But, again, to hell with it—I had a nine o'clock class in calculus and the sailing was a little rough, about then, in the intricacies of differential equations. And so, I forgot all about it.

Of all the curiosity seekers, helpers, and enthusiasts who were constantly swarming around during the course of our commercial and experimental diving operations as the school years progressed, there appeared one young man who stood head and shoulders in every way above all of them—Prescott Brown. He was diving with me every week, and it looked very much as if here was a pal who was going to stick through the rigors of this crazy business for a long time.

Prescott was definitely the inventive type. Brainchildren of his were everywhere apparent. His car was bristling with gadgets of his own design. He had even made a diving helmet, similar in principle to Jack Browne's but beautifully built from a workmanship standpoint.

We were planning to spend a few days in New York, partly on some diving business matters and partly just to have a good time in the big city. It was now December, and the roads were in the worst possible shape for our two-hundred-and-fifty-mile night drive. But this didn't seem to worry Prescott very much—in fact, he seemed rather delighted at the chance to test his various self-invented gadgets in his car.

Prescott was very thorough, a nice characteristic for diving work. As we prepared to leave Boston, he gave his car a last check, like a locomotive engineer leaving for a long run. Lifting up the hood, I noticed that a light automatically went on like that in a modern refrigerator. The engine itself and the entire interior of the "engine room" were finished in a spotless white glossy enamel and were as clean and inviting as a Sears Roebuck model kitchen. Because of the iced highways, he had slapped his dual wheels on the rear, an attachment of his invention, with chains covering both tires. Operated from the driver's seat was a chute that deposited sand in front of each of the four wheels with a pull of the lever, to cancel the terror of high-speed stops on ice.

All the way down the main route between Boston and New York, I didn't see a car traveling over fifteen miles per hour on the iced highways. However, Prescott's car, more like a high-speed tank than an automobile, whizzed down the icy roads at a speed that made them look by comparison like parked cars.

It started to snow a little as we entered Connecticut, which couldn't make the roads any slipperier but did seriously impede the visibility. We were now in some sort of a city and were completely lost. It was late at night, and anyone who hadn't sense enough to be in bed by that time was at least staying inside. We drove and drove, desperately trying to peer out of the segments of windshield cleaned by the wipers to pick up some trace of human life as a possible source of information.

The schooner *Seth Parker.*

From my window, I noticed the vague silhouette of what appeared to be a ship. I rolled down the window quickly. There, looming up, was the hulk of a huge schooner.

"Holy smokes, look—for God's sake, hold it!" I yelled, and Prescott brought his begadgeted Ford to the stop that only his car could make on a night like that.

We got out in the blizzard and walked along the dock to which the massive vessel was tied. Her four giant masts and shrouds towered heavenward and disappeared into a ceiling of snow pouring out of the inky black night.

I had never before seen one of those giant windjammers, which today have practically been replaced by the modern steamship. We walked aft, thrilling as we almost brushed against her unpainted weather-beaten planking. To us, this type of ship, more than anything afloat, symbolized the romance of the sea. Soon we were at the stern, and our eyes strained upward trying to catch a glimpse of her name through the flurry of snow. There it was, stretched across her transom—the *Seth Parker!*

Adventure-minded Prescott had read hungrily about the Seth Parker Expedition, too, and here we were, by pure accident, almost brushing the very ship.

Excitedly, we walked forward to go over our previous casual observations more carefully on this famous vessel, which we had thought was merely a cargo schooner; every portion of her immediately assumed an even added glamour.

Midships, we noticed the gangplank leading up from the dock on which we were standing to her main deck far above our heads. A chain was hung across its lower end to which was attached a terse statement of policy, "No Visitors!" We looked at the sign, looked at each other, and without a word being spoken, were as wholeheartedly agreed on what we were about to do as if we had come all of the way from Boston to try just this.

For our long legs (Prescott also being six-foot-three), that chain was just a good step. A moment later, emerging on her expansive deck, we saw that our explorations would probably be unheeded. There was no sign of light or life on any part of the ship. We walked forward, tracking a light film of snow that had fallen on her deck, and gazed at each new sight. Her huge masts towered upward beyond our range of visibility. We studied and gingerly touched the massive booms and parts of the intricate maze of standing and running rigging. On the forward deck, we carefully examined her head stays, winches, bow sprit, and anchors.

Thus far, we had been unopposed—the worst that could happen, we silently reasoned, would be that we would be kicked off, and so we decided to explore the interior, also. Finding no doors locked, under the forward deck we soon found ourselves studying the galley, mess hall, and rope lockers. We were deterred only when we heard the sound of a half-dozen sets of snores as we entered the fo'c's'le (forecastle) head. We left there, however, only after we had sneaked a look into this portion of the ship, which was the bunk room for the crew.

Walking aft, we peered down into the hold and then into her engine room. Two huge power plants were visible. I remembered I had read that Phillips Lord had put in twin screw diesels so that he could safely maneuver her into many of the strange uncharted waters he planned to visit on his round-the-world cruise.

The night was getting late, and we still had quite a few miles to drive to get into New York. We decided to wind up our explorations with a visit to her after cabin. Through the passageway, we saw to our left the glass-paneled radio room where, apparently, the broadcasts would be sent out. And then—the final treat—the after cabin!

This room, in marked contrast to the rough unpainted timbers of the rest of the ship, was luxuriously finished in a beautifully grained natural walnut. Red leather-covered and overstuffed pieces of furniture were individually bolted down to the deck, since this ship was going to sea. On the port wall was a glass-fronted case with an elaborate array of guns, from revolvers to huge elephant rifles. Our eyes studied the many interesting objects, from the ship's bell to the old-fashioned melodeon fastened down in the corner, which, I remembered, Seth loved to play. We were veritably tingling with excitement at the thought of the adventure ahead for this ambitious expedition.

Suddenly, by some strange instinct, I sensed that someone was staring at me. I wheeled around, a pang of fear shooting through my body.

A grim, expressionless figure was staring piercingly at me. Blue-green eyes peering out of a tough weather-beaten face watched our every move. He was sitting in the cabin corner and very possibly had been there, motionless, ever since we entered the room.

"Hello," I said, frightened to death. I stuttered for words. "We were driving—going from Boston to New York—got lost—don't even know where we are—suppose this is Bridgeport."

He just glared. Why, for God's sake, couldn't he say something?

"—just driving along the dock," I stumbled on. "Saw the schooner—I have never seen a four-master before—then noticed she was the *Seth Parker*."

He still glared. His face was as cold and expressionless as that of a corpse.

"We had both read of your expedition—there was nothing to keep us from coming on board—except, of course, that little sign, 'No Visitors.'"

His eyes warmed. His face softened. He laughed. "Sit down," he said.

I went on to explain that we were both adventure-crazy, that I had gone into professional deep-sea diving, and that I owned and had built quite a bit of underwater equipment.

We were anxious, not to talk about our interests but to hear about the Seth Parker Expedition. However, our questions were invariably quickly and briefly answered, and we were right back to diving again. I told him about some of our experiments with underwater cameras and lights. I told him about the powered diving bell I had built.

His name was Flink—Captain Constantin Flink—and he was an Estonian. He was getting very talkative, now that we had told our stories.

He should never drink coffee in the evening, he told us. He had taken several cups at dinner that night and couldn't go to sleep. He had tired of tossing in his bunk and had come into the after cabin just to sit and think. He had been there when we came in.

He had followed the sea ever since he was a boy in the old country. His seamanship had been the kind before the mast, of the "iron men and wooden ships" variety.

Now he was navigator of the *Seth Parker* under Captain Phillips Lord. He tactfully admitted that his was the responsibility for the schooner's safe navigation. Captain Lord was captain in title only (for publicity purposes) but was not too familiar with the intricacies of the deep sea.

Not only had his hostility vanished, but he was now very friendly. He went on to tell us, to any degree in which we were interested, of the forthcoming expedition.

They would go down the Atlantic coast to Miami, then east through the Bahamas, the West Indies, and down that long string of lesser Antilles to Trinidad. On the cabin wall was a chart of the world, and his finger continued to trace out the course, skirting South America to Buenos Aires and thence across the South Atlantic to Cape Town. Then, Madagascar, India, Indochina, the South Sea Islands, and through the Panama Canal to New York.

He paused in his itinerary to tell us about the Death Ruby of the Cambodian jungle, where expedition after expedition had been mysteriously wiped out by an alleged curse on whosoever should seek that stone of fabulous value. It was buried under a mammoth four-headed Buddha, marking the remains of a lost city, in the almost impenetrable jungle of Indochina. But Mr. Lord would get it.

They would stop at Formosa, where roamed a biologically mysterious lizard resembling an artist's most fantastic conception of a dragon. They would investigate the Human Tree of Madagascar, said to ensnare and devour unwary human beings.

They would swim in the blue lagoons of coral atolls and, at night, dream to the strumming of the native guitars in the Islands of Paradise.

They would hunt tiger in India and elephant in the African Belt.

It almost seemed like a dream. But here we were, sitting in the cabin of this very vessel, which was now actually started on that fantastic adventure. At our very fingertips were the massive redwood timbers of that ship.

What an experience this had been for us! Flink, probably, would have been glad to have anyone to talk to in these lonesome hours of the sleepless night. It was now time for us to be rolling on to New York—it would be daylight before we could get there as it was.

All of the time, naturally, I couldn't help but imagine myself on that expedition. A number of things that Captain Flink had said about public interest in it had convinced me that even a thought in this direction was just a waste of energy. Furthermore, it was just a few weeks before final exams at MIT, and that was really my first responsibility.

But just for the hell of it, I thought that I would ask. He had been so grand that I almost hated to risk such a trite question.

"What, Captain Flink, would the chances be of signing on the expedition?" I blurted out as we were standing up preparing to leave.

His face lit up. "You remember you read that Mr. Lord was planning to do a little diving for old pirate treasure?" I remembered very well. "Well, he has been on the lookout for someone who knows something about diving and who has his own equipment."

I blew all of my fuses.

He went on to explain. There had been quite a number of professional divers who had applied for the job, but he had refused because they could be of no value to the expedition other than as divers. After all, the diving would be just on occasions. Whoever took that possible post would also have to be able to do some other job simultaneously. The entire crew had been completely made up six months ago with no changes to date—and there were forty thousand applications that had since come in.

"Did you say forty thousand applications?" I asked, unable to conceive of how small my chance would be. No more fuses were burning out at this point.

"But," he said, his eyes twinkling in the dimly lit cabin, "I'll give you a little tip!"

His eyes searched me. "You say you have had three years of mechanical engineering at your Massachusetts Institute of Technology—you should know something about engines. Mr. Lord told me yesterday that he is planning to fire the assistant chief engineer—he's no damn good!"

"Gosh!" was all I could say.

"You say you are going to New York. You can find him at the offices of the National Broadcasting Company, 711 Fifth Avenue, New York City."

I tried to thank him, but all that he could say in his own inimitable brusque manner was, "Get going—I'm getting sleepy now."

We covered the remaining miles between Bridgeport and New York in record time, arriving at the big city as the glow of the rising sun started to fill the sky. There wasn't time for much except to get cleaned up, have breakfast, and make a few hasty plans for the presentation of my application to radio star Lord.

On the split second of 9 a.m., Naval Observatory Time, I arrived at the offices of the Seth Parker Expedition, a group of rooms that had been granted to them by the National Broadcasting Company.

"Do you have an appointment?" lethargically asked the young woman whose business it apparently was to act as outer guard between a celebrity (who was polling top honor in radio popularity throughout the nation) and an overzealous public.

"Captain Flink sent me down here to see Mr. Lord," I explained. I intended to take full advantage of his voluntary tip at least to get by this firm and efficient first barrier.

"Where and when did you see Captain Flink," she asked, still visibly unmoved.

"I spent most of last night with him on the schooner lying at Bridgeport," I replied as firmly as I could. "It wasn't over four hours ago that I left the ship to come down here to see Mr. Lord."

She insisted on knowing what the nature of the business might be, but I was just as determined not to tell her. If the phrase "looking for a berth on the expedition" ever slipped out, I was certain that I would never get in.

Finally, she broke down and said, "all right," and I made my first big step of progress. I moved from the first outer office to the second outer office.

There were nicer chairs to sit on in this more sacred inner sanctum; not having closed my eyes all night, this was no minor consideration. Also, the woman in charge of this room was more cordial.

This was Thursday morning. Thursday noon arrived, and another woman came in to relieve my contemporary pal (and we were good friends by now) so that she could go out to lunch. There was no one to relieve me, and so I just continued to sit.

People were rushing back and forth, telephones were ringing, and many signs of activity were going on, but none of them seemed to affect me. An occasional assurance would come back that Mr. Lord had been notified that I was waiting to see him but that I would have to wait. The afternoon, as long a stretch as I ever remembered, dragged on, and at five o'clock, it was apparent that numerous of the staff were leaving. Perhaps my chance would come after everyone else had left—then I could not only see him but would catch him a little more away from the pressure of the height of the day.

Six o'clock came, and my stomach felt as if it were full of a flock of flapping butterflies. However, my pain was soon to be relieved. A woman came in and advised me that Mr. Lord had left for the day. She was "so sorry" that he had been too busy to see me.

The next morning, Friday, at nine o'clock, I arrived and assumed my same chair, the contour of which was by now very familiar to that section of my anatomy which had learned it so well yesterday. Two incidents broke that day's sit.

The first was an actual glimpse of Lord. Someone had simultaneously left two doors open and, for a fleeting second before they were closed, I saw his face two rooms away—a face with which I was now well familiar from many hours' study of a picture on the wall.

The second incident occurred when I asked, probing into expedition business as much as I could, what those piles and piles of bales were leaning against the opposite wall almost up to the ceiling of the room. "Those," the woman explained with a searching look in her eye, strongly suspecting that I was looking for a berth, "are applications for jobs on the expedition. Those are just the early ones—we don't keep them anymore—there isn't room."

Five o'clock came, and soon after, I was told that the big boss had left for the day.

Saturday was just like any weekday at the office. By Saturday night, I sensed that the woman was beginning to feel a little sorry for me. I told her that I would have to leave for Boston on Sunday since I had some quizzes on Monday that I couldn't cut. So, she gave me a little tip.

"The offices are closed on Sunday," she said. "However, quite often Mr. Lord comes down on Sunday morning to get a little work in without interruption. There will be no one else in the office. He usually leaves the door unlocked. You might just walk in and, if you can talk fast enough, might be able to edge in what you have to say before he kicks you out."

My very low hopes jumped again.

Sunday morning, I arrived a little late, as she had suggested, but found the door locked. Knocking brought no responses. Listening, ear to the door, revealed no sign of human habitation.

There was no place to sit now—I was just out in the hall. Wearying of standing, I decided to go out and come back. I had something to eat and then returned to find no further sign of life. Again and again I tried it, until the last deadline approached. Prescott had occupied himself with other things during the past three days, but today we definitely had to leave. It was now about 12:30 p.m. We had decided that we would start at 1:00 p.m. for Boston, since we both had a few things that had to be done Sunday night. I still had to get ready for that Monday quiz.

I pounded on the office door for the last time and decided that Fate just hadn't meant for me to meet Mr. Lord. Even if I did, I only had a tiny chance of getting that berth. And so, to hell with it—I certainly had tried.

I was broken-hearted, as I went down the elevator, to leave my hope unaccomplished here in the middle of the fourth day.

The elevator opened into the main lobby of the building and I, the solitary passenger on this Sunday morning, stepped out. A man with hat pulled down over his face brushed by me and stepped in. For some reason, I wheeled around quickly, sensing something familiar about him, which as yet I couldn't see. He stepped into the car, turned around to face the front, and in a split second as the door almost closed upon me, I stepped back in, too.

It was Phillips Lord.

Here we were—all alone in an elevator. Here was my golden opportunity.

"I'm Max Nohl, Mr. Lord—I've been trying for four days to see you."

"What do you want?" he said impatiently with the same voice I well remembered from his radio programs.

That elevator was shooting upward like a rocket, and I knew that our forced confinement in this cage would be very short-lived, since his floor would be the first and last stop. So, in fast phrases that I had mentally rehearsed for the past four days, I told him my story—Captain Flink, my diving experiences and equipment, the assistant chief engineer who was going to be fired, and my talents at assistant chief engineering.

His first interruption was, "With what type of diesel engines are you familiar—air injection or solid injection?"

My God, I thought, *are there TWO kinds?* But I was really rejoicing. It seemed to be down to a fifty-fifty chance—whether I could guess the right one. I imagined that he, as suggested by Flink, didn't know anything, either, about diesel engines. I had to pick one.

"Air injection," I said boldly, and before I could add that I was also familiar with solid injection, just to try to cover everything, he said, "Good—that's what I have on the *Parker*."

We were now entering his office, and he excused himself for a second, leaving me in the familiar room where I had spent the past three days. He came out in a few seconds, tucking some papers into a briefcase, which I imagined were copy for the next radio broadcast.

He was now genuinely friendly. "I must go out and catch a bite to eat—why don't you come along and we'll talk it over?"

Ten minutes later, his quick lunch was finished, and I was a member of the Seth Parker Expedition.

I was to join the ship in ten days. This would give me time to go back to MIT and see what arrangements I could make with my professors and the dean about final examinations (which were only a few weeks away, but for which I would not have time to wait); it would give me time to get packed up and buy the few things I would need on the ship, and—I hoped—it would give me time to find out what in the hell an air injection engine was.

I might have lied a little about my knowledge of diesels. Perhaps it wasn't a lie at all, strictly speaking—we had had, in theoretical thermodynamics, a brief contact with the "heat and entropy analysis of the perfect cycle of a compression ignition engine," which, someone had whispered, was the diesel. However, I had never seen a diesel engine nor did I understand how one worked. There was a course in diesels that I had planned to take next year.

However, I had ten days between now and the time that I should come face-to-face with the *Parker*'s diesels.

Back in Boston, I went up to the library and obtained every book available on diesel engines. I haunted every professor and instructor who taught diesel engineering. I went out into the laboratory and studied the engines that MIT used for practical instruction. Also, Prescott's father, it turned out, owned a diesel cabin cruiser and probably knew as much about them from a practical standpoint as any of the faculty.

Those were a hectic ten days. As finally I left Boston to join the ship in Philadelphia, where she had put in for her weekly Tuesday night broadcast over NBC, I had felt that if there was anything about diesel engines that I didn't know, it either wasn't known or wasn't worth knowing. One disadvantage I had had was that I didn't know the make of the engines the *Parker* carried, and so I had the additional burden of familiarizing myself with all makes.

MIT had been very generous about my courses. They had arranged to give me early exams in a few of them, for which I was prepared; another few more were given to me as a present on the basis of my term's work; and the remaining few could wait until my return from the expedition.

I arranged to arrive in Philadelphia in the early evening so that I could sneak belowdecks that night and look the engines over. I had my "diesel bible" with me, listing practical information about every known make. I could look at the nameplate of my forthcoming charges and then go back to my cabin and spend as much of the night as was necessary reviewing every detail of that particular model.

Upon my arrival at the schooner, Captain Flink was very cordial and congratulated me on selling my bill of goods to Lord. He introduced me to the rest of the gang, who were all very friendly, and I had a warm feeling that everything was going to be all right.

Flink assigned me a cabin midships, which, except for Mr. Lord's stateroom, was as nice as anything on the ship. I signed on, and was rather pleased to note, with everyone else except Flink and the chief engineer getting five dollars a month, I was to receive fifteen a month. Mr. Lord had previously told me this, which I interpreted at the time as an act of generosity, since I, and I imagine any of the other forty thousand applicants, would have fought just as hard for the chance to go without any pay at all, just to go. This small fee, he had explained, was just to give us a little something for cigarettes and spending money.

Then, after everyone else had turned in, I made my first trip down into the engine room.

I had been soaking in diesels for about ten days, but there was something about these that looked lots worse than anything I had read about. They were huge things, probably standing eight feet high and bedecked with a mess of piping, tubing, rods, and gadgets the likes of which I had never dreamed existed.

But first of all—the nameplate. It said "Southward Harris." I couldn't remember any such name. I brought out the book that supposedly listed every diesel ever made. There was no such name.

This upset my plans a bit; but, after all, it was a diesel. I was supposed to understand diesels now, so I decided to figure it out for myself.

Wearing my new overall work pants, I started tracing the myriads of tubes and pipes to find out which went where and what all of the multitude of levers, valves, and gadgets were supposed to control.

Pipes disappeared into sheaths of other pipes down into the bilge or into inaccessible parts of the cylinder block. Some I could trace, but I could see that, to a large extent, the chief would have to familiarize me with the details of this maze. After all, what could he expect.

The next morning, I met Whiskers. He was the chief engineer.

His nickname was a mild description. It was questionable whether a razor or scissors had approached his face for a good many years. His head was just a ball of hair, with kindly laughing blue eyes beaming out and a place in the shrubbery opening up when he talked.

All of my worries vanished into thin air. Whiskers was a regular guy. His real name, seldom heard by him, was John Parks.

After morning mess, we lingered over the table and talked. He told me of some of his experiences at sea before the mast—he was a genuine schooner man as well as a very practical sort of diesel engineer. I told him about some of our diving experiences. We talked about diesels, at which time I assumed that he was trying, in a very friendly way, to sound me out on my knowledge of them. This was a very gratifying experience, since I soon realized that I knew a lot more than he did about the theory of combustion ignition and diesel design. I was beginning to wonder how he could handle them as well as he must be able to, with as little basic understanding of the physics involved. Any doubts in my mind soon turned into a tremendous admiration as we stepped down into the engine room, and he painstakingly went over the practical end of operating and servicing those twin giants.

Whiskers and I were always the best of friends. I resolved, even though his education had been extremely limited, never to show any but the highest respect and goodwill toward him, even after I learned all the practical details of handling the equipment.

Although I had read that Lord's crew was to be made up of "college students," I found that I was the only one on board who had ever stuck his nose inside the so-called advanced halls of learning. On the other hand, I knew less about the figurative and literal ropes of the ship than did any of the more experienced professional followers of the sea who made up the crew.

At breakfast that first morning, I was the butt of the conversation. Flink, in his booming voice, trained to outride the heaviest offshore gales in ordering his men, asked for what degree I was working.

"Bachelor of Science," I told him.

"BS," he guffawed. He laughed so violently, I thought that he was going to strangle himself on his breakfast. "Max Gene Nohl," he roared, almost shaking the timbers of the ship in his delight. "You know what BS stands for!"

"But, Captain Flink," I interrupted, "you are wrong. In Latin, the degree for which I am working is known as *scientiae baccalaureus*, and at MIT the degree is given like this: Max Gene Nohl, SB."

Flink roared louder than ever. The whole crew was laughing with him. As soon as he could control himself, he announced, "That's better still. I know the name we'll have for you.

"Did you ever go shark fishing?" he went on. "You know what they use for bait? They buy an old dead horse, which is good and dead, and they let him rot for several weeks. Let him rot until he stinks so bad that you can't get leeward of even a hunk of that stuff. The worse it stinks, the better the bait it is. Then you put it on a hook, and if it stinks bad enough you catch a shark right away.

"You go down in your diving suit. Maybe that's all you are good for, too. Max Gene Nohl, SB," he roared again. "From now on, we'll call you Shark Bait."

12

DISILLUSIONMENT

For many years, I had heard about Phillips Lord. It was almost impossible for anyone who paged through newspapers or twirled radio dials not to run across his name or one of his numerous radio programs occasionally.

I had never particularly followed him, or any other radio program for that matter, in the struggle of getting through MIT. However, from the snatches I had picked up, I could not help but like and respect the beloved character he portrayed—Seth Parker.

Millions of Americans had followed his Sunday evening meetings over the National Broadcasting Company, almost to the point of worship, as he gave forth his meaty bits of homespun philosophy and religion, always in the spirit of his oft-replicated expression, "You go to your church and I'll go to mine." Perhaps to those uncounted millions, he was bringing, with his countryside philosophy and humor, something more real and living than they sometimes found in their own more formal houses of worship; perhaps to others, shunners of the church, he was bringing the essence of righteous thinking in a fashion so subtle and yet so captivating that they little recognized the real spiritual want which he was satisfying.

What a wonderful experience this would be—to live for the next few years with as fine a man as this, on an adventure-seeking expedition that would certainly take us through Paradise and Hell.

All I knew about the expedition was what I had read in the newspapers and magazines. From those colorful accounts, the expedition itself would certainly be as fine and as sincerely devoted to the spirit of adventure as the man himself.

"Captain" Phillips Lord on the schooner *Seth Parker.*

My first feeling that the expedition was not as well organized as the glowing press reports had indicated came on my first day of work. I had just descended into the engine room to officially meet my two twin charges, the giant power plants of the *Parker*.

In our first few minutes down there, Whiskers told me the story. "They call it twin screw diesel in the newspapers," he said, "but they are nothing more than two big piles of junk."

He went on to explain that our two were the first diesel engines ever built in the United States and were never intended to be more than a pair of experimental units. Lord had bought them cheap, but no matter how little he paid for them, if it were only one dollar apiece, he had been swindled.

This was a little in contrast to the press releases and stories describing the fine power plants we had to take this vessel into the treacherous uncharted waters of the four corners of the earth.

"The engines," Whiskers continued, as he pointed out the various features, "are not only not worth a damn but they were put in backward."

The port engine was put in on the starboard side and the starboard side on the port. The port propeller was on the starboard shaft and the starboard propeller on the port shaft.

"Thus, when the signal comes 'full speed ahead!' you set them on 'full speed astern,'" he explained. "When the telegraph clanks 'full speed astern' set them on 'full speed ahead.'" Thus, they would always be running backward during normal forward motion of the ship.

"By running backward and by having the open side of the crankcase inboard—i.e., facing each other—they are continuously splashing oil on each other," he continued, as I could well see—they were just dripping with black sludge.

"They'll throw fifty gallons of oil in an eight-hour run. It all goes down into the bilges, and the whole bottom of the ship is by now a little lake of lubricating oil—at twenty-five cents per quart or whatever he pays for it."

The worst part, however, was that the injectors, built before they knew how to design a diesel engine, forced the oil in a concentrated stream directly down on the upcoming piston, almost like an acetylene cutting torch. In ten to fifteen hours of running, that streak of flame would melt right through the piston.

"Then," he grinned, "is when you are going to see hell break loose. The flame shoots down and ignites the crankcase, and a sheet of flame goes out at the other engine that is enough to scare ten years' growth out of anyone. Then all we have to do is put the fire out, shut her off, and then overhaul; and that overhaul, Mr. Shark Bait, is what we are going to be doing most of the time—even if we only use the engines for emergency."

He showed me an entire shelf of extra pistons and then pointed out a welding outfit with which we could try to patch them up.

I had arrived in time to be present at the broadcast in Philadelphia. There was much excitement as everyone watched the second hand of the wall clock and awaited the signal that we were on the air.

"We are broadcasting from the fo'c's'le of this old schooner," said Lord in his best radio voice. I didn't quite understand. We weren't in the fo'c's'le—that was the extreme forward part of the ship. This was the after cabin—the extreme stern. This was not supposed to be a fictional story—this was the real story of a real expedition. The entire value and substance of the broadcast was that it was a description of events exactly as they were happening.

However, I assumed that this was all part of the ethics of the trade. We were actually now in show business, and anything that we could do to heighten the illusion of reality was presumably to be within the license of the art.

Down the coast we went, each Tuesday night putting in to another city to lay a wire from the ship to the nearest telephone line for feeding the program to New York, from where it would be set to the many stations of the National Broadcasting Company.

Almost every Tuesday night would come the familiar part of the program, "We will now transfer you to a microphone on deck," and then after the proper interval, in would come his voice, "It's a beautiful night up here on deck—the stars—the reflection in the water." But never did any microphone venture out of the after cabin.

At the end of each program, we were supposed to be hoisting sail for our journey to the next port. "Stand by on the mizzen halyards!" intonated Captain Lord, as one of us, often I, pulled a short piece of rope through an old creaking block hanging in the after cabin, which was doctored up for a good squeaky sound effect.

I helped with other of the sound effects. The deep-throated foghorns, adding harbor atmosphere, were not really passing steamships, but just one of us, at a prearranged cue, blowing into a long wooden whistle borrowed from the NBC sound effects department.

Because I had the gift of being able to speak articulate English, I soon found myself in the capacity of a barker, rather than an engineer, during the days when we were in port. For the sum of twenty-five cents, any Seth Parker fan could board and visit the vessel. We would take them, in groups of six to twelve, through the parts respectively assigned to us and point out the items of interest.

Business was good. In some of the cities that we visited, the lines were blocks long, four abreast, waiting to pay their quarter to take the tour through this widely publicized adventure-bound schooner.

During the course of the tour, the visitors were given the privilege of purchasing personally autographed photographs of Lord in character as

Seth Parker. They could also buy (for $2.50) an autographed Bible or an autographed Seth Parker hymnal.

These were genuine signatures—I know because I autographed several thousand of them myself. One of the seamen and I could duplicate Lord's signature almost indistinguishably from the real thing.

By the time we reached Miami, the last castles of illusions had been dashed to the ground. However, we had had a lot of fun, and I found that the gang on board was solid gold—particularly Captain Flink. Behind his hard, merciless exterior was a warm heart, a clear head, and forty years of experience at sea. We worshiped the man. The other captain, Phillips Lord, was not faring so well in our esteem. However, we had seen little of him, since he had been on the ship only on Tuesday nights for the broadcasts, commuting back and forth by plane or train each week from New York as we proceeded down the coast.

In the morning, we were to leave Miami and the United States. However, it was a disheartened and badly swindled crew that was still sticking. Hearts formerly light with the anticipation of adventure ahead were now gone or in the depths of gloom.

Worst of all, our sponsor, Frigidaire, suddenly refused to renew its contract. We had no explanation given to us as to why they had discontinued their plans for what we thought would be two solid years of weekly broadcasts from the ship. It certainly was not because the programs were not successful, for we found public interest in this expedition almost fanatical everywhere we went. Anyway, whatever the reason, we were suddenly and unexpectedly without a sponsor.

According to the press releases, this expedition was supposedly the fulfillment of a lifelong dream of this great radio star, to be paid for out of the fortune that he had made on the air. In actuality, it was intended to be a profitable business venture. But now, with no sponsor, we were on the other side of the ledger—it was no longer a moneymaking but a money-spending trip.

Lord, even when the money was coming in, had tried so hard to economize on everything and conditions had been so bad on board that, one by one, most of the professional members of the crew had quit. Flink alone was being well paid and was sticking. "What in the hell do I care?" he would say. Whiskers was sticking. I was sticking. I had my diving bell, several diving suits, a half-ton diving pump, a lot of miscellaneous underwater cameras, open helmets, etc. I couldn't quit—how could I ever afford to ship all of that stuff home? Anyway, I wanted to see what was going to happen.

The rest of the crew, a fine bunch of boys, were more notable for their good nature than for their experiences at sea. A Boy Scout with a merit badge for cooking was now our chief cook. We had no assistant cook. There was not a single AB (able-bodied seaman—the lowest papers awarded to professional sailors in the Merchant Marine) left. We had just our four cadets left, all boys of around high school age who had come down the coast with her. That was all!

However, at the very last minute the next morning, we bribed one of the ABs to come back, and we sailed.

Captain Flink, as well as numerous other professional sea captains, had been arguing from the very beginning that "you can't sail around the world to the east!" All the way on the radio star's projected route, Flink said, we would be bucking a headwind and stemming a strong current. "It can't be done," he insisted.

But Lord had been taught that there is no such word as *can't*, and we started east.

Whiskers had gone to a machine shop and had forty-eight extra pistons made for those diesels; we would be very careful to save them for only extreme emergencies. At sailing time, who arrived at the dock to fill out our depleted crew but four beautiful young women—Hollywood movie starlets—who were going with us, too.

We sailed and sailed and sailed. We were heading for Bimini, a British Island in the Bahamas, only fifty miles from Miami. We could see it, and we would sail all day, but it would not seem to get any closer. Finally, a freak beam wind picked up, and we went into Bimini on a port reach. The difficulty with the *Parker* was that she had the hydrodynamic lines of a big wooden box. She was meant to be a slow-moving freight schooner for Pacific Island service and was a very poor ship to windward. Now, practically with no cargo, she sat on top of the water like a cork and did not have enough lateral area below the water line to allow us to sail into the wind.

Bimini, our first foreign port, was the first real feeling of expedition that we had. From there, we headed south and east through the Bahamas and West Indies.

On and on we sailed, frantically bucking headwinds and currents. We weren't getting far, but as far as we were concerned, we were getting a lot of sailing, like a dog learning to run on a treadmill.

Lord was now on board, which added a tension that we had not known on our trip down the coast. He was extremely irritable, ordered us around, and swore at us. Little cliques were forming. Everywhere the boys were airing their gripes to one another.

One night, our AB, a professional drinker as well as a professional sailor, decided that he personally would "kill that #?&#?# #?¢@& Captain Lord." He went out into the galley, his inhibitions well paralyzed with rum, and grabbed a huge butcher knife, more adequate for slaying elephants than human beings.

Lord had stepped into the mess hall and the inebriated AB rushed at him in a violent lunge. If it had gone home, that blade would have driven not only through him but practically through the massive hull of the ship.

I developed a new respect for Lord when I saw him handle this emergency. Displaying no sign of fear, he stood calmly, posed apparently for no defense. Then as the knife approached him, up swung his left arm to harmlessly divert the blow, and out lunged his right square to the sailor's jaw. He hit that man so hard that he literally lifted him right off of his feet, deposited him on the mess table, and slid him its entire length. A few huge bowls of hot food reposing in the middle of the table for the night watch went over like tenpins. He was a sorry-looking sailor now, as unconscious as a dead herring and soaked from head to foot with sticky thick soup and mushy beans.

"Put him in his bunk," said Lord tersely with no visible sign of emotion.

The feeling against the expedition was growing. The boys deserted the ship like a bunch of rats the minute our anchor was down. This problem was a matter of growing concern to Lord, since we were not allowed to sail from foreign islands until we had all hands on deck—those small governments did not want to take the responsibility and expense of getting American sailors back to their homes. A few members of the crew would invariably attempt to drown their feelings under tremendous quantities of native rum, which, they found to their delight, could be purchased for practically nothing. Ten cents in American money and a bit of sharp trading and they could get as oiled up as a month's ordinary pay would allow in

New York City. The bottle itself was usually worth more than its contents. They would gather up a few empties on the ship if they were short of cash, and they soon learned that two of them were a fair trade for a single full one. These boys would simply disappear for days. The *Parker* could not sail until they had been found, usually in some, to put it mildly, unpretentious local dive.

At Cape Haitien, Haiti, Lord decided that he was not going to go through this again. He anchored approximately three miles off the point, planning to take the motor launch himself into the town, but forbidding anyone else to go ashore. It was Sunday, and it had always been understood that on this day, unless we were standing at sea, we were free to do what we liked.

Our new cook, whom we had picked up someplace along the way and who was one of the violent dissenters of the expedition, hadn't heard the bad news. He had fixed up some cold grub and laid it out in case anyone stayed on board and got hungry, but he, like everyone else, was planning to go ashore and do what most sailors do when they go ashore. He had been busy in the galley when the announcement came, and somehow or other, no one had told him. Thus, having finished the breakfast chores, he went to his locker and crawled into his brand-new blue serge suit. We had been at sea for some time, and today, only the sky would be the limit. In recognition of how very short Sundays can be after long weeks at sea, he had started drinking at breakfast, burrowing into a special stock of native rum that he had bought up at the last island.

What type of entertainment he had in mind was unsaid, but it must have been something extra special to warrant that brand-new blue serge suit.

Emerging on deck, he scanned the disheartened-looking gang. "What's wrong with you guys, aren't you going ashore?"

Someone volunteered that Mr. Lord had ordered, "No shore leaves today."

"The hell you say?" he spouted, noting our confirming nods.

With not a moment's hesitation, he went over to the shoreward rail, climbed clumsily up, stood on it, and turned his head to grinningly gur-gle, "Have fun, fellas." Over he jumped, blue serge and all, in a thirty-foot belly flopper that would have slapped a more sober sailor unconscious. We were just three miles offshore, three miles of boiling breakers and seething sharks, to the rocky coast of Haiti. There wasn't much that we could do to help him. Lord had gone ashore, and all the boats were locked up.

He made it! We saw him three days later—the Haitian police brought him out in one of their government launches. He had rolled down into the bilge of their boat, apparently as lifeless as a sack of flour—not from the swim, but from three subsequent days of soaking in the native rum. That blue serge suit! Apparently, he had torn it to rags in making a landing on the jagged rocky coast through the heavy white surf. But there he was, our cook, all in one piece.

This broke the ice—nothing on earth could have kept the rest of the gang from going ashore, too, during the many days we lay to. Cape Haitien offered many interesting short expeditions.

Tension and unrest, however, were growing almost to the breaking point. Shore leaves helped us forget our troubles, but back on ship, we found ourselves as part of an expedition that was on a very unhappy and unstable basis. No one much cared what he did—Lord had lost the good-will and support of his own crew, and his increased bitterness toward us was only building up the vicious cycle.

We rounded Haiti and put in to its capital city, Port-au-Prince. It looked very much as if this would be the end of the Seth Parker Expedition. No sponsors had appeared for the broadcast. The experimental short-wave transmissions back to New York had been so noisy with static and accompanied by so much fading that they were unsatisfactory for regular rebroadcast. Admiral Byrd, just recently, had successfully used a similar system from the North Pole, which was undoubtedly the stimulation for this expedition—but Byrd, we now learned, had more elaborate equipment for multiple-wave transmission, so that instantaneous change could be made from one band to another without interruption. We had only one band, and if that clogged up or faded, that was the end of the broadcast.

Understanding now that this expedition was in reality more of a busi-ness enterprise than the fulfillment of a lifelong dream, I actually felt sym-pathetic toward Lord's plight. Our Frigidaire check was no longer coming. There were no more fans who would stand in long lines for hours to con-tribute a quarter to see the ship. There was little hope for a new sponsor because of the technical shortwave difficulties.

Rations had been cut. The colorful exciting adventures that we were going to have were materializing as just a lot of hard work. We were dangerously shorthanded, and everyone was putting in extralong hours, but the compensation, which was supposed to be just a token in addition to a large dish of real adventure, was still only five dollars a month for most of the crew.

There is probably something about any ship at sea that tends to breed discontent. The discipline is traditionally rigid. The quarters are confined. The comforts and joys of home are absent. There are no dates, shows, sports, hobbies, or entertainments. The constant rolling and pitching is tiring and keys up nerves.

The spirit of forthcoming adventure could offset all of this. But now, there was little hope, and it was no wonder that discontent was running rampant through the ship.

We never did find out how our words of discontent got back to Lord. Some of the boys suspected hidden microphones in the forespeak, where bitter gripe sessions constantly brewed. Others felt that right in our midst was a stool pigeon, squealing back to Lord, but who could it be? Everyone seemed to be on our side.

Whatever it was, the very walls of the ship seemed to have ears. Our gripes and plots were leaking back to Lord—word for word. We were learning to hold our tongues until such time as we were ashore.

Every one of the boys was fed up with the whole thing. We had all looked forward to being members of an expedition in which we would all share alike in work, play, adventure, or come what may. Yet, we were not members of the expedition. We were flunkies. Lord was stingy with the food, although we had countless tons of good things to eat down in the hold. These were not things that he had bought but out-and-out gifts that had been presented to him by respective manufacturers in return for advertising testimonials. Most of it was deemed too good for the crew, and we fared, by his order, on the cheap stuff.

We never had the privilege of using any of the expedition equipment. One day, when off duty, I asked Lord if he had any objection to my trying out one of the outboard motors on which I had been working. We must have had a dozen of them on board, all advertising presents. He came back with the heartwarming, "Who the hell do you think you are? You're here to work and not ride around in an outboard motor!"

Since he had been on the ship, from the time we sailed from Miami,

After mess on the fore hatch, with Nohl's diving shell at left. COURTESY OF KATHY END

his contempt toward us had been almost unbelievable. We were in quite a strange set of circumstances; we couldn't quit, and he wouldn't fire us. The island governments repeatedly warned that they would not allow him to sail without all hands on board unless Lord posted for each man ashore maintenance until such time as the next ship would depart for the United States and then left with the government sufficient funds for passage back to New York. They were not going to support any American sailors in addition to their present manifold charitable problems.

With a fare of approximately $150 back to New York, each man was a substantial liability to worry about. Whether we departed of his or our wishes was of little consequence to the islanders.

With most of the boys there was more than just a desire to get back—the desire for vengeance was deep.

Lord certainly was well aware of the high feelings against him. From the trickling of our discussions back to him through the mysterious ears, it was apparent that he was extremely anxious to protect himself from whatever was brewing.

13

Departure

Although I later had to admit that I accomplished nothing spectacular in my total diving activities to the entire expedition, I did feel as though I had enjoyed a wealth of wonderful experience that would broaden my background in my chosen profession.

On one occasion, the *Parker*'s propellers got fouled up in a wire rope. We had tried frantically, by alternately running the engines forward and in reverse like a car rocking out of a deep rut, to break the blades loose but just succeeded in binding them tighter. With no facilities for hauling out a ship of that size in those tropical islands, a diving job was the only possible solution. Sitting on the giant propeller blades in my diving suit, after many exhausting awkward hours, I was able to cut through countless strands of the taut cables and finally free the ship.

On numerous occasions, I had opportunity to go down into really deep water with the diving shell. This was deluxe diving, not one extra ounce of pressure bearing down upon my body as I dropped down to depths of two and three hundred feet. The large window afforded wonderful vision. We rigged a little fin on the after-portion of the shell so that she could be towed slowly from the ship such that the window would always be facing forward. With the two-way telephone, I could instantly call up and ask to be raised should some large formation suddenly loom up in front of me.

After I had finished my time on the bottom, usually determined by the waning patience of the boys up topsides rather than by any desire on my part to leave this beautiful and fascinating world, I marveled at the speed at which I could return to the surface. Since there was no pressure on my

Nohl with his diving shell, ready for a test dip. COURTESY OF KATHY END

body, no decompression was required, and they were free to haul me up as fast as the little deck engine would wind in the wet cable.

As interesting and comfortable as was the diving in this air-conditioned pressure-free shell, I was constantly being reminded of the limitations of this type of equipment. At frequent intervals, I wanted to pick up a sponge here or a piece of coral there and found myself completely frustrated. That same heavy metal casting that was so perfectly staving off the tremendous external water pressure was also serving as an impenetrable barrier, making it impossible for me to put my hands out into the water. I was locked inside of this powerful steel bubble with no conceivable function other than to peer out of a tiny window. Even this was, at times, extremely trying, since as the schooner rose and fell with the surface swell, so did the diving shell. I had no control of the direction of my window, and it always seemed that what I wanted to look at was just out or just turning out of the narrow cone of my restricted vision. The two functions, both available to the rubber-suited diver—dexterity and maneuverability—were both totally missing. It was, in a sense, like riding in a train and gazing out of a tiny round window—I had merely to look at and enjoy what was passing by my window and not be concerned about what I wasn't seeing. I couldn't touch anything. I couldn't pick up anything.

In a little machine shop maintained by the *Parker* for servicing her power plants, I soon built up a small underwater camera, which housed a simple two-dollar Brownie. Although the pictures that came from it were photographically far from startling, I was learning the nature of the basic problems that I would have to solve in subsequent years of real professional underwater motion picture filming. Although I was well familiar with the theory and practice of air photography, the rules didn't quite seem to apply down here. However, I was as proud as a brand-new father with my little Brownie snapshots showing a drab but blurred little fish swimming by or just a solitary little piece of staghorn coral standing there.

I had learned that Davy Jones is not as cordial and gracious as his lavishly beautiful submarine panoramas might indicate. I had tried very gingerly to pick up a sea urchin and found him as friendly as a porcupine. For days, I and the amateur surgeons in the crew were picking out the countless festering little barbs that had sunk deep into my hand. A little stingray, which I had not seen, had given me a nasty slash with his barbed tail. I had cut myself badly many times on the beautifully delicate and multicolored coral formations, which I had found were as sharp as a hardened-steel cutting tool. I had seen many a little octopus—and a few that weren't so little—scamper away at my approach and was already recalling with much suspicion some of the harrowing tales of fiction I had read of attack by these evil-looking creatures.

On Sundays and evenings when we had been allowed shore leave, I had frequently forsaken my associates in the native cabarets and sought out more of the local fishermen. With my trickle of Spanish and their occasional better trickles of English, I soon found myself gathering some further very interesting statistics on wrecks throughout the tropical islands. Some of them were old Spanish, British, and French galleons with alleged fabulous fortunes in their rotting timbers; others were more recent and more modern vessels carrying possibly less fabulous but probably more practical opportunities for salvage. I made careful notes on every detail. Someday, I would come back with my own expedition!

I suddenly realized that I had now been living on this ship for over half of a year. There was little hope ahead for the Seth Parker Expedition. I found that I was spending my idle hours dreaming of my own expedition. I was anxious to leave.

I was receiving fifteen dollars a month. Checking at the steamship office in Port-au-Prince, a rapid calculation soon revealed that even if I saved every penny of my salary, it would be more than a year before I could accumulate my fare back to New York. It was obvious that I must find some other way.

A few days later, one of the boys came down into the engine room and said that Mr. Lord wanted to see me.

Several minutes later, in his office, he explained that he soon would have need for some diving equipment but also realized that I was anxious to get back to the States and that the future of the expedition looked rather discouraging. The shortwave broadcasts had not been getting through to New York without considerable noise and constant fading. No sponsor appeared to be interested in the program, probably because of these technical difficulties. Without a sponsor, the expedition could not continue.

If I would rent him my diving equipment, he would guarantee its safe return to New York. As rental payment, he would buy my first-class passage back to New York. Was I interested?

"It's a deal!" I exclaimed.

The SS *Columbia* was due into Port-au-Prince at seven o'clock the following morning. She would sail at noon for New York.

I went below to my cabin and packed all of my personal things. I went over all possible questionable points about the diving gear with Jack Love, one of the boys who was going to stay and who had shown the most aptitude and interest in the equipment. Before turning in that night, everything was ready for my departure.

At 7 a.m. the next day, bedecked in my practically unworn white flannels, white shoes, and nautical blue tailored jacket, I stood at the rail and watched the beautiful white ship glide into the dock about a mile from us as we lay at our anchor. As I veritably tingled with excitement at the sight, Mr. Lord came up on deck.

"What in the hell are you doing?" he sharply asked. I reminded him that he had told me that I was leaving at noon on the *Columbia*, and I pointed to her as she was now being made fast to the dock.

"She doesn't sail until noon," he curtly stated. "Get back to work!"

I resurrected my discarded work clothes, which I had thought I would never see again, and went back down into the engine room to tinker for a few hours with the dilapidated diesels. At ten o'clock, after not hearing the expected call, I emerged and again reported to Lord, "I had better start getting the grease off of me—the ship sails in two hours."

"Go back to work!" he ordered.

At eleven o'clock, I again appeared before him, feeling that this was the last possible minute.

"Go back to work!" he again ordered. "I'll call you in time."

Eleven thirty came, and I felt that it was time to take matters into my own hands. Down in the engine room, I cleaned as best as I could the grease off of my hands and slipped into my spotless clean clothes. At 11:45, I again emerged on deck. The *Columbia* was blowing her whistle.

Lord came up and saw me. He didn't speak to me but called to one of the men and asked him to report to his office. They both disappeared below.

At 11:55, the *Columbia* again blew her whistle. I could see her men casting off the first of the dock lines and preparations being made to raise the gangplank.

He never had any intention of letting me go, I thought from obvious conclusions.

The last hawser was cast off, and I could see the huge white ship ease away from the dock. She was now underway. My last hope was collapsing.

Suddenly, the man who had been conferring with Lord appeared on deck. "Let's go!" he called. "We'll have to hurry."

The boys leaped to the call and helped get my things into the speedboat. A minute later, we were churning up a white mountain of water as we raced toward the moving ship with engines wide open.

We overtook the *Columbia* in the outer harbor where she was gracefully moving along at half speed. Our speedboat came up alongside and, with skilled handling, kept exactly abreast. A moment later, they lowered

what I later found out was a cattle sling from a swinging cargo
boom. I awkwardly managed to
get in with all of my belongings
and then heard high overhead
the order, "Hoist away!"

With a jerk, up I started. I
called back, "So long!" through
the mesh of the net as I saw the
boat beneath me race up her
engines and turn out.

Up the great steel wall I went,
over the rail, and in a moment
felt myself and my belongings
being deposited on deck with a
thud.

A husky, tough-looking
officer came up to me. From
the insignia on his beaten cap, I
could see that he was the bos'n.

"Who are you?" he gruffly asked.

"I'm Max Nohl," I replied. "I'm from the schooner *Seth Parker*. Arrangements have been made for my first-class ticket to New York."

He roared with laughter. He looked me over, noticing my clean white
shoes and white flannels. He laughed again.

"So, you're going first class to New York?" He was laughing so hard he
could hardly talk. "Brother, here's a chisel and here's a mallet—get out of
those clothes. You're going to work. You can start by chipping paint on the
port well deck rail."

There was no mistake about it. I had no ticket back to New York. Mr.
Lord *had* spoken to the captain. He had arranged that I *work* my way back.

Actually, I enjoyed the trip more than I would have as a first-class passenger. I didn't have a bunk and had to sleep on the mercilessly hard steel
deck. I didn't have any work clothes and had to borrow some discarded
overall pants, several inches shy of reaching my ankles. Most of the first-
class passengers came "slumming" to see the comic effect created by my

lanky form in the tiny clothes and didn't hesitate to laugh at me to amuse their immaculately attired ladies. However, I learned much about the running of a modern ship at sea and made many friends among the members of the Puerto Rican crew.

The prospect of landing in New York, a huge but merciless metropolis, flat broke was not too pleasant to contemplate. However, I could ward off starvation for quite a number of days. We were carrying a load of bananas. I found some empty boxes and packed my clothes into these. Into the empty suitcases and duffel bags went all of the bananas I could carry.

Although I must admit that for a year afterward I could not look a banana in the face, I soon found myself home, and the Seth Parker Expedition was a past chapter in my life.

I still was much interested in what Lord was going to do to wind up his highly publicized venture. He could not just quit. Not Phillips Lord. He was too much of a showman. Before I had left, we had often envisioned the headlines that we guessed he was probably planning. We could vividly see them blazing across the newspapers of the country. "Seth Hero of Shipwreck" or "Lord Endangers Own Life to Save Crew as 'Parker' Sinks."

The rest of the story I learned second- and thirdhand. I wasn't there.

Lord finally gave in to the cold fact that if the expedition was to continue, it must sail west. The *Parker* could not buck the headwinds and head currents that it would encounter on Lord's determined plan to sail around the world to the east.

Leaving Port-au-Prince, he hoped by going on that he might pick up a new sponsor. The next stop was in Jamaica. They were trying to shoot some possibly revenue-producing motion pictures, one scene of which was to be taken at night by flares. The entire pile exploded unexpectedly, and Whiskers was burned so badly that there was grave question as to whether he would live. He was sent to New York, leaving the engine room staff almost depleted.

On went the *Parker*. Through the Panama Canal she was towed. Ahead now lay only the vast wastes of the huge Pacific Ocean. As I understood, he next planned to deviate slightly before venturing out into the sea-lanes of the Pacific and visit the glamorous Galapagos Islands. Actually, these islands are barren, rocky, and very much lacking in glamour to a would-be inhabitant. The hungry, jagged, sharp submarine rocks could, however, quickly spell the end to the expedition—an end in what was popularly a very romantic corner of the earth.

Whether or not this was Lord's intention, I do not know. Some of the boys in the crew thought so. Whatever might have been the purpose in reaching the Galapagos Islands, the *Parker* never made it.

As was true throughout the expedition, her blunt coffin-shaped hull floated high on the water like a cork, and she was unable to sail into the wind. It was necessary for her to fight a head tide and headwind to reach those islands. She sailed and sailed and sailed, at one time actually coming within sight of the bleak rocks, but was never able to reach them. Gradually losing against the relentless elements, the *Parker* finally found herself being swept to sea.

Caught in the equatorial currents and trade winds without engines, on they drifted and sailed. Week after week, month after month slipped by. The Pacific Ocean seemed endless. Suddenly, flashing through the ether came the most dramatic of all radio calls, "SOS." The *Seth Parker* was in distress several hundred miles off Pago-Pago, American Samoa. The now almost forgotten expedition loomed in the public print once again.

Unfortunately, the nearest ship to her was the HMS *Australia*, a heavy cruiser bound from Australia to England carrying on board no less a personage than the Duke of Kent. The *Australia*, true to the traditions of the sea, turned back in her course several hundred miles to offer assistance to the "stricken vessel."

However, the *Seth Parker* was not stricken. She was boarded by the officers of the Admiralty warship, inspected, and a radio report issued that the SOS had been a fake. She was in as seaworthy conditions as she could be. She was dry and her canvas and rigging were intact, and so the *Australia* departed.

Hours later, Phillips Lord, not to be rebuffed by the mere British Navy, sent out a second SOS. Again, unfortunately, the *Australia* was the nearest vessel to receive the message, and again, in traditional style, she was required to heed this sacred signal. A second time, she steamed up to the *Parker*, her officers boarded and inspected the schooner, and the radio message was sent out that Lord had tried a fake SOS a second time.

The American press had Seth back on the front page from coast to coast. The evening paper in my own hometown, the *Milwaukee Journal*, ran a two-column front-page box, headed by "Britannia Rules the Waves, but Seth Parker Rules the South Seas."

And then, to top it off, hours after the *Australia* had left the *Parker* the second time, out through the ether went the third determined attempt— SOS! Again, the *Australia*, still the nearest radio-equipped vessel, was required to turn back. What happened to warrant that third SOS, I don't know, but by the time they arrived, some damage had occurred to parts of her rigging. The seriousness of it can be estimated by the fact that this time the British cruiser took some of the men on board but steamed away immediately, leaving the *Parker* with eleven men still on board. That was the end of the Seth Parker Expedition. Several days later a tug came out from Samoa and towed her in.

The *Parker* was sold and converted into a fishing vessel. My diving equipment was all sold. Not a penny of the price ever came back to me.

I had lost my diving bell. I had lost all of my diving equipment. I had spent all of my savings before the expedition had ever sailed in an effort to have my equipment in the best possible shape and to be fitted with every possible accessory that I would need. It was now all gone—money, equipment, and half a year of my life.

I went back to MIT. My father was willing to finance the balance of my education.

Vose Greenough was not using his equipment and was willing to loan me what I needed. I had found some secondhand stores in Boston and New York where I could pick up some occasional bargains. It was not long before I had a complete rig again and could go back into professional diving for professional fees.

Vose finally decided to sell his gear and, with the money I now was earning using it, I paid him in full. I now owed nothing, was solvent, had all of the basic equipment that I needed, and would soon finish my courses so that I could devote full time to my chosen profession.

I was once again finding myself dreaming of my own expedition.

BOOK III

THE *JOHN DWIGHT*

14

THE *JOHN DWIGHT* MURDERS

"Three hundred and fifty thousand dollars! Think of it—it's all ours if we want to go down and get it!"

Thus spoke Dave Curney. His eyes, through his thick bifocals, bulged out at me like those of the goggly-eyed fish friends with whom he had worked for more than twoscore years of professional deep-sea diving. "I can drop a lead line plunk on her deck," he ranted on to allay my apparent suspicion as to the probability of our finding the wreck. "Think of it— twenty fathoms of water—three hundred and fifty grand—we can't miss!"

It was a cold bleak day on Martha's Vineyard. The frigid winter winds sweeping in over the vast icy waters of the North Atlantic seemed to chill the very core of this little Massachusetts island on which we were standing.

My friend Vose Greenough had taken me down to meet the veteran diver Dave Curney. Dave knew the exact location of the sunken steamship *John Dwight*. Vose had been quite certain that Dave would listen eagerly to any sort of a proposition that would make possible the salvage of her precious cargo. Vose had been right—Dave was very much interested!

A couple more years had gone by since my return from the Seth Parker Expedition when I had gone back to MIT to resume my studies. I had picked up quite a few more pieces of diving equipment. With the help of a few college friends, I had built up several underwater cameras, a battery of deep-sea lights, a new experimental diving suit, and numerous, novel open helmets. I had found many occasions to test and use the new equipment. There had been job after job, such as groping for the body of a woman who had committed suicide in a quarry, reshackling harbor mooring lines, inspecting the long piles which support many of Boston's buildings

Diving on the wreck of the *JM Allmendinger* with one of Nohl's first underwater cameras in Lake Michigan in July 1934. Left to right: Jack Browne, Verne Netzow, and Max Nohl.

constructed over open water, caulking submarine pipe, and retrieving anything that had been lost in the water that was valuable enough to warrant the expense of my "professional service." During the intervening summers, Vose and I had located and profitably salvaged a number of salable items from several lost steamers in the Great Lakes but had found no fabulous fortunes in their rotting hulls.

In a few more months, I would be graduated from MIT and would not again be hampered with the confinements of my college courses.

I had bought a boat. Rising and falling with each tide in a muddy slip in South Boston was my first love, the salvage sloop *Silver Heels*. She was a rough but unbelievably rugged vessel, measuring thirty-eight feet on deck and fifteen feet in beam, the latter giving her a rare stability for handling heavy loads over her side. She had a large, heavy-duty marine engine for power and carried a tremendous spread of canvas—a large gaff rigged mains'l, gaff tops'l, stays'l, and an enormous spinnaker.

Thus now, talking to Dave, I had bargaining power. Dave knew where the wreck was and had probably forty years of professional diving experience to contribute. I had a salvage vessel, crew, complete diving rigs, and the necessary cash to cover operating expenses. What more could we need?

"It's a deal!" exploded Dave.

I was to be graduated from Tech in February 1935 (only the sixth year of my four-year course—with expedition interruptions), and we would sail as soon as the North Atlantic had passed off the worst rigors of her stormy winters—perhaps about April 1. In the meantime, we would complete the changes that would transform the fishing boat the *Silver Heels* had been into the salvage vessel that we envisioned. This would mean many refinements in living and working accommodations on board ship; the addition of air compressors, power hoists, elaborate diving and salvage gear; plus, a complete overhaul.

In the meantime, also, I would find out more about the *John Dwight*, the mysterious murder ship for which we were shortly to search. From the morgues of many newspaper files, from interviews with some of the old Coast Guard men who had been stationed in her waters, from tale after tale by amateur sleuths who had studied the almost fictionally fantastic trail of clues, and from Dave who had lived and breathed that wreck for all of these years in anticipation of an opportunity to recover her treasure, I gleaned the following story.

The *John Dwight* was a small steamship which, for many years, had been used as a coal lighter. During World War I, the United States Navy had used her, but at the cessation of hostilities, she had been recommissioned, according to her papers, "for towing and wrecking."

During the winter of 1922–23 she lay in Newport (Rhode Island) Harbor with no apparent sign of life. Suddenly, early in the spring of 1923, to the surprise of local waterfront wiseacres, she was hastily fitted out and suddenly went to sea without apparent adequate preparation.

She next officially showed up in the harbor of Vineyard Haven, the city on the island of Martha's Vineyard, which was Dave Curney's home. However, there had been a considerable lapse of unaccounted-for time between her Newport departure and the Vineyard Haven docking.

Dave, always interested in every waterfront activity, had by quirk of circumstances visited the *Dwight* and had, probably without invitation, peered down in the hold to view her cargo of "Canadian flour"—flour in one-hundred-pound cloth bags and also in wooden barrels. This, according to the manifest, was her cargo. To Dave, on that inquisitive visit, there had been something very unreal about that reported cargo of flour.

Her last port before Vineyard Haven, on which entry we could get very little information, had been Montreal—she had been bound from Canada to the United States. This was the year of 1923—Prohibition at its peak in this country. Dave was very suspicious that she carried more than flour.

The next series of events came over the gangster grapevine—there was no official report on them, but the newspapers ferreted out the following.

The next leg of her journey was back to her home port of Newport. She had dropped anchor in the outer harbor after dark and her dual skippers, Captain King and Captain Carmichael, had moved ashore in one of the lifeboats. They contacted a group of their friends and made arrangements to sell the contraband cargo. As was the custom in bootleg circles, a substantial down payment was made on the spot with the understanding that the liquor would be picked up just before daybreak the following morning. A ten-dollar bill had been torn jaggedly in two, Messrs. King and Carmichael retaining their half and the shore group taking the remainder. The matching of these two sections would serve as proof at delivery time as to the rightful owner.

The main bulk of the cargo was in Black & White scotch whiskey with a wholesale value at the dock of approximately $350,000. In addition to this, there was a substantial quantity of Frontenac ale and an undetermined number of cases of rye whiskey of unknown make. The rye had been loaded in first, followed in turn by the ale and scotch, and then covered by only a sufficient number of flour bags and barrels to deter a casual observer. Even some of the wooden barrels were filled with miscellaneous bottles well sodden down and packed with the white bleached flour.

As a deposit toward the agreed price of $350,000, the buyers had given Messrs. King and Carmichael $100,000 in cash. The two captains had returned to their ship with this in their pockets plus their half of the torn ten-dollar bill.

It was a dark, foggy night in Newport Harbor, and as they boarded the deck of their ship, they suddenly were inspired. Scarcely able to see their own bow as they peered forward from the after quarter, they suddenly realized how simple it would be to slip out through the dark wall of mist and disappear into a black nothing to miss their early morning rendezvous. They had a hundred grand in greenbacks in their pockets—and they still had the scotch.

They were next sighted in Providence the following night. Apparently, they planned to do the same thing over again, but they would have to act quickly before the underworld telegraph spread out the warning signals. Here, we understand, King and Carmichael rowed in, arranged to again sell their illegal cargo, and again slipped out into the dark of night. This time they did a little better, collecting $125,000 as a down payment, giving them a total of $225,000 in cash—and they still had the liquor.

This was a new racket in the bootlegging business and, so far, had every indication of yielding better profits than the more conservative and old-fashioned method of just buying, selling, and delivering the contraband cargoes. However, these players of an illegal game didn't believe in any rules, and they failed to recognize that they now had two strikes against them.

No one knows the next destination of the *John Dwight*, if it was to be in this world, since she never made it.

It was early the following morning, April 6, 1923, that the veteran Coast Guardsman Roland Snow was standing his regular early watch, vainly peering out into the pea soup fog from the lookout tower built above the Cuttyhunk Coast Guard Station. Cuttyhunk is the tip island of the Elizabeth group stretching seaward south of Massachusetts's Cape Cod, and this bleak outpost provided a vigilant watch over the steam of ship traffic between Boston and the north and New York and the south.

Probably Snow was wondering why a man must spend half the night in a glass room looking out into a fog so dense he could scarcely see the rocky ground of the island beneath him.

And then suddenly, bellowing over the water came the roar of a terrific explosion. It had come from the open sea and must be that of a ship. But other than to note this, there was little that observer Snow could do. Launching their surf boats, the Coast Guard would have no even approximate direction in which to steer nor an idea for what to look. He could only watch and wait. Approximately five minutes later, what well might have been a miracle occurred. The fog lifted in almost a tunnel-shaped formation for a few seconds and allowed the hawklike eyes of Snow to see the bows of the SS *John Dwight* slipping beneath the waves of the Atlantic Ocean. In those few seconds, he took a compass bearing on the wreck, and immediately the always-ready surfboats were launched into the water to rush aid to her survivors. Pulling through the again-settled mists, they arrived at the scene of the disaster only to find a few scattered pieces of floating wreckage—but not a trace of a human being.

One other clue appeared a few days later. The freighter *Dorchester* had been feeling her way through that same fog as she sought for the channel of Vineyard Sound. Entered in her log for that early morning watch was the notation that her lookout had seen three men in a lifeboat pulling for shore. Surprised that such a boat would be at sea at that early morning hour under the low visibility conditions, he had called to offer aid, but the men in the almost overrun small boat refused to reveal their faces and pulled frantically away from the *Dorchester* for shore.

The following day, April 7, fishermen found the bodies of seven men floating in Vineyard Sound. All had been wearing life jackets and thus did not sink.

The subsequent autopsy revealed that three of them had been murdered before they went overboard, for there was no trace of water in their lungs. The skulls of all of them had been brutally bashed in from behind, and their throats had been cut. Some of them had been badly mutilated with the apparent motive of destroying any possibility of identification.

We believed that the *Dwight* had probably been headed for Boston or another north and east port to resell her cargo again, but that the boys in the crew had not quite shared either in the enthusiasm or the profits of the new "down payment racket," and that there had been a mutiny between the two captains and the crew.

Rum runners customarily carried a substantial dynamite charge in the most tender portion of their bilge, prepared for instantaneous discharge from the helm with the logical thought that it would be far preferable to totally lose their ship and its cargo of evidence than be caught by a revenue cutter. If the *Dwight* was so fitted out, it is very probable that some of the boys in the crew, in the throes of the mutiny, detonated this charge and tore open a gaping hole in her bottom, preferring possible escape by boat to an inevitable meeting of the two parties who had been looking for $225,000 in down payments and who knew no form of retribution other than machine-gun bullets.

Apparently, just one lifeboat was launched as she went down—and in that boat were the two officers of the ship, Captains King and Carmichael. The eight boys in the crew who were still alive had to swim for it as the deck sank beneath their feet. The bodies of the other seven, undoubtedly killed during the mutiny, were still in the ship.

As we constructed the story, the two captains rowed about to the struggling sailors in the water and offered them in this hour of mutual desperation a friendly word as offer of assistance. As they, one by one, attempted to climb over the gunwales of the boat, down over the backs of their heads came the terrific wallop of an oaken oar handle crushing the tender posterior portion of their skulls and killing them instantly. There would be no cries to warn the others; the dull thud would not be heard over a few feet above the roar of the seas; the black blanket of fog would veil from sight each individual atrocity.

King and Carmichael escaped. Their boat was found beached high and dry, and the tracks of two men were discernable in the sand leading from it.

Tucked under the thwarts of that boat was the grimmest of all the remaining evidence as to the ruthlessness of these men—there was the body of John King, the son of Captain King. He had been murdered in the lifeboat.

15

MAN OF IRON

"**M**y God, I made it!"

The marks had just come in on my last set of final exams. I had passed them all—not with any very impressive margin—but I now had that part of my life out of the way. I had my SB degree—and now, we could go to sea.

Three Milwaukee boys were going in with me on the forthcoming expedition.

Truman Marsh was now a veteran of deep-sea sailing. A few years before, he had been invited to be one of three men who were planning to take a thirty-eight-foot yawl out of San Francisco to try and hit a tiny pinpoint of land jutting out of the water four thousand miles away—Tahiti.

They had bought a sextant, a chronometer, and a book titled *Celestial Navigation*. None of them had ever tried this art before, but they followed its instructions and shot the sun as they stood on the dock in Frisco. They sailed the next morning and hit Tahiti "right on the nose"—their first sight of land in four thousand miles. Truman had just come back from this journey, and now, after a few months ashore, jumped at my invitation to sail with us on the *Silver Heels*.

"Uncle Hal," as he affectionately was known to all of his Milwaukee friends (Harold Stein on his driver's license) was to be engineer. Uncle Hal was a wizard on engines, on anything mechanical, or just on anything.

"Doc" (Norman Hoffman, of Milwaukee) was to be—well, we hadn't quite decided. He was a member of the expedition and would help wherever needed. We knew we could always count on Doc, however tough the going might be, to be on hand and to help as best as he could, although he had no particular seagoing talent.

Then, of course, there would be Dave Curney, with whom I had the fifty-fifty contract on the proceeds of the *John Dwight* salvage. Dave was a strange character.

We were wondering, as we heard story after story of his past life and were now beginning to see him in action, if it ever would be possible for anything to kill him. He was a very short, extremely heavy-set man whose body was as resilient as steel. He had the muscles of a giant, covering an extremely heavily boned frame. He was truly a "man of iron," and although he was a good foot less in height than most of us, we and everyone else were afraid of him—afraid because there was nothing that would stop him except probably a .45 bullet. Dave was a heavy drinker, like most divers, and he was a dangerous man when drunk. Coca-Cola bottles would actually smash on his head and apparently not faze him in the least. He had many times been clubbed with a timber that would crush the skull of any normal man, but the wood would break on his head seemingly without his noticing it.

Dave had been in numerous automobile wrecks—some of them head-on collisions—but he would walk away from the accidents. He had once fallen off his diving raft into thirty feet of water with his entire suit on with the exception of the helmet, and thirty-two minutes later, he had been retrieved, pumped out, and gone on with the dive. He had fallen out of mastheads and second-story windows—essentially anything that a drunk could fall out of—and had never even hurt himself. People had long since learned not to fight with him. He didn't even bother to guard himself—his opponents, assured knock-out blows, would do no more than damage or break their fists. Dave would then let go a right that had almost killed many a man.

The boat had needed a lot more work than we had anticipated. But we knew it would be impossible for us to sail on our scheduled April 1, although, essentially, we had the bulk of our fitting outfit completed.

The rising sun had barely advanced its forerunning glow of light into the Atlantic sky on the morning of May 29 when we were suddenly wakened by a terrific commotion on deck. "Get the hell out of your bunks.

The *Silver Heels* in her South Boston slip. Left to right: Max Nohl, Dave Curney, Truman Marsh, and Harold Stein.

What are you guys goin' to do—sleep all day?" bellowed out a booming voice. The one-man commotion turned out to be Dave, as he lumbered clumsily through the small cabin passageway of *Silver Heels*.

It was so early that we were in doubt whether this was a late-last-night or an early-this-morning visit, but it soon became apparent that it was both. Dave had obviously been drinking, for in a few minutes, the fumes from his breath turned the cabin air into almost an explosive mixture. He swore on in a violent blue streak that almost seemed to shake the massive beams of the boat.

As soon as he had worn to exhaustion his alcohol-inspired boyish delight in waking us up, his voice quieted down to its more normal level of a mere roar. It seemed that he had four days off—tomorrow was Memorial Day, he had arranged to take the following day, and then came the usual weekend. Dave was diving on a nearby government job.

We were well stocked with food and water. Fundamentally, almost everything really was ready. There was more painting and fancy trimming to be done, to be sure, but this chance might not come again for a long time—perhaps never. We had to have Dave to show us where the wreck was, so we quickly ran over in our minds how well we could fare with this ship as she was. There were extra gas tanks that would need mounting before we could use them. We needed some new charts, but we could get by on the very old ones that had come with the boat, even though they were not as detailed as we wished. We would have Dave, after all—the seer who spurned such "sissy aids," the sage who had such an uncanny knowledge of those waters and such an ingrained instinct of the sea that we ourselves had almost begun to believe that the man was the god that he claimed to be.

We all felt that it was a mistake to go with anything less than perfect preparation, but at the same time, here was opportunity knocking.

The engine was taken down, but in a few hours Uncle Hal could have it together and in good shape. Nothing had been tested; but then what better way could there be to test our boat, equipment, and crew than on a shakedown cruise?

At almost exactly twelve o'clock noon, we hoisted anchor and felt the boat take on way to move for the first time out of our muddy slip into the fast-moving waters of the channel as we picked up the ebbing tide.

It was not over two hundred yards from our anchorage that we were almost to meet disaster.

We had several bridges to go through. At the first, the attendants were obviously eating lunch, and so we cautiously approached the low structure to the side of the main channel, staying out of the swirling current. Finally, after catching another bite, the operator went to his controls, and we saw the huge mass of steel swing open. I was at the wheel and threw the clutch in to guide our vessel past the concrete abutments, our forward motion plus that of the current carrying us through at almost express-train speed.

A few hundred yards ahead was the next bridge, where the attendants were also leisurely eating lunch. And then, suddenly, the motor revved up, and I found that the clutch had disengaged. I pushed frantically at the lever but could not get her to bite.

Truman, seeing our plight, grabbed the horn and started blowing frantically in the small hope that the next bridge would open before we reached it. We were hurtling along at a terrific pace, but without power, we had no steerage. The wheel was as useless as if we had neglected to fit the ship with a rudder.

I could see the attendant leisurely walking over to the control house but now knew that there was no chance of its opening in time. In only a few more seconds, we would crash into the low span, and a moment thereafter, we would emerge on the other side in a mass of splinters. The collision would certainly tear out our mast and pull up the deck, and the heaving rigging (and firmly bolted chain plates) would instantly rip the entire upper body of our boat. Uncle Hal made an inspired leap for the engine room, grabbed a wrench, and without even a thought, veritably dove at the clutch adjustment. With a quick turn, he readjusted it for actuation by the cockpit lever, and just as we all were about to close our eyes and duck our heads for the forthcoming crunching of our craft, she shuddered and stopped short as the energy of the racing motor reached the propeller through the reverse gear.

Our next incident happened about a half an hour later. I was still at the wheel, steering the *Silver Heels* along at full cruising speed through the rather narrow steamship channels leading out of Boston Harbor. Bearing down upon us was the huge hulk of a fast steamship moving apparently at full speed. They were off our starboard bow, and by every elementary law of the sea—the rules of the road—they had the right of way. However, they were changing their course to meet the channel, and I soon found that I would have to deviate considerably on a right rudder to avoid them. Perhaps I waited too long or did not quite anticipate their change of course, but soon I found that we were getting uncomfortably close. With a fast turn, I threw the helm hard over, but after the first pickup of the rudder resistance, suddenly the wheel spun freely. I frantically lifted the lid to the gear box and saw that the pin connecting the rudder shaft to its quadrant was sheared off. We had no steerage.

Truman, always quick on the trigger in an emergency, knew just what to do. We had two lanyards cautiously connected to the upper after edge of our rudder. These were to prevent its loss in the event that it should accidentally slip out of its well at sea. With our diving rigs, we could always reinsert it. Truman leaped to the stern, grabbed the starboard lanyard, and with a terrific yank, pulled the rudder hard over. Quick to respond to

Max Nohl (left) and Dave Curney, man of iron (right).

her helm, our head swung immediately over to take us out of danger. It was now but a few minutes' work to fashion a new shear pin, and soon we were underway again.

Dave had been ranting about all morning long, swearing a blue streak at every detail that held up our sailing. By the time we were actually moving, he was sound asleep, flat on his back, sprawled over an anchor and numerous galvanized iron deck fittings with as little concern as if they had been feather pillows. Through our bridge and shearpin incidents, he had not so much as missed a single snore.

But now, with a start, he awakened and leaped to his feet. "By God, we're at sea!" he cried sarcastically and lumbered aft to the cockpit. Uncle Hal was watching his charge, the engine, like a mother, oiling and reoiling everything that moved. Doc was taking a nap; Truman was either dozing or studying some papers below; and I was ready to lie down for a little while. We had worked hard all morning after a brief night's sleep, shortened so considerably from both ends that not much had remained. Accordingly, when Dave demanded to take the wheel, I was glad to be relieved and decided to go below for a nap. I gave him the course I had been steering, "South by west a quarter west." He let out a blue streak of profanity that I was too tired to clearly hear as I disappeared through the companionway.

It was probably several hours later that I awoke with a warm glow of satisfaction as I felt my bunk heave up and down beneath me to the rhythm of the oncoming seas, felt the faithful vibrations of our engine, and heard the water running furiously along the hull. As I lay there, I found myself mentally comparing the motion of the boat to that when I had left the wheel to Dave. It seemed as if she were pitching considerably more but rolling less, indicating that we were probably running cross seas. Perhaps the wind had changed, but we had been running in the troughs before—almost parallel to them.

I went up on deck and land was no place to be seen. A glance at the compass and it was evident we were heading due east.

"What's your course, Dave?" I asked him meekly, wondering who was crazy, he or I.

Dave unleashed a streak of violent oaths such as I had never heard. I just stood there and watched. A few minutes later we were going northeast. Suddenly, he threw her head over, and we were heading due north. Then after a few more minutes over, she went the other way, and we were pointed east by south.

Dave was steering by some strange instinct not available to most men—and apparently it was not working that day. We had long known that his vision was very poor; he wore bifocals that appeared to be almost spheres of glass held on a frame before his eyes that magnified them to a giant's size.

Apparently, for several hours we had been either heading to sea or madly circling. The man must be blind. We had often before suspected that Dave was only able to discern shapes. He could read, we had noticed, but the print would actually almost touch the end of his nose as he perused occasional matters of interest to him, the paper moving back and forth by his battered beak like the carriage of a typewriter.

Dave was proud—so proud that he would not even now admit that he wasn't on his psychically sensed course. Nor did we tell him.

Truman and I sat down at the chart, and we tried to calculate the most probable position in which we might be from the meager data with which we had to work. Our position, to state it mildly, was indeterminate. Another complication was readily apparent—we had a very small gasoline supply on board, and this wild ride must have put a bad dent into it. The wind, what there had been of it, was rapidly falling off.

We hoisted every square inch of canvas we owned, but soon found our sails flapping in the limp air. We still had a chance of hitting our first destination—the gas dock at the mouth of the Cape Cod Canal—before running out. What else, we thought, could we do but try?

Doc cooked up a mess of grub, and we all gathered in the cockpit. Against the blazing red sky of the setting sun, we enjoyed our first meal at sea on the *Silver Heels*. *This is the life*, we all thought, as we scudded along over the darkening Atlantic.

Our meditations were suddenly interrupted by a scream from Dave. The devil was in him again—it was either that or, as we more strongly suspected, the contents of several bottles he had snuck on board. He was like a crazy man. I never thought that such a concentration of profanity could be crowded into one curse. It was not clear as to what was aggravating him, but his violence was of such proportion that we all suddenly seemed almost petrified with fear. He was ranting as if in a delirious nightmare, something about how he was going to kill.

Suddenly, out shot his fist with a wallop that could perhaps shear a man's head from his body. He struck at the companion door. His iron fist ripped through the pine boarding with a crunching and splintering of wood that made us shiver to our very toes. His arm went through the half-inch panel up to his elbow.

"You don't want to fool with Dave Curney," he stated simply, now silently and deliberately eyeing each one of us in turn with a glance that made us decide that from now on, with him on board, we would not sleep without a watch.

Our watches went on through the night. A low mist had dropped in over the water, which destroyed any possible chance of picking up a light and identifying it for position on our charts. When I turned in for some sleep, Uncle Hal was at the wheel and Doc was running occasional soundings at the bow. Sooner or later we would probably approach some portion of Cape Cod, and from a study of the charts, we could determine our approximate distance from the beach by the depth of water.

Doc was singing away in true seamanlike style, "By the deep, thirty-three," "by the mark, thirty," etc.

Sometime after midnight, Uncle Hal came below, woke me, and asked, "What shall we do? We're in a pea soup fog." Up on deck a moment later, I found that he had under-described it. Doc was still running constant soundings, and the water was now shoaling off. Hal was running at reduced speed, veritably feeling his way along—and that was all that we could do.

About 3 a.m., the engine coughed, picked up, coughed again several times, and then stopped. We were out of gas. From the soundings, we knew that we were close to the Cape or some land. The engine had been running for fifteen hours. Now to have it quiet seemed almost as much a shock as to find oneself suddenly naked on the street. It even woke Dave up, who had again been sleeping over the anchor on our forward deck, for suddenly out of the darkness boomed his voice, "Breakers ahead!"

If he couldn't see, he certainly could hear. We strained our ears and soon knew that he was right. Dead ahead through eight points (90 degrees) of our bow was the roar of the surf.

We dropped our first anchor in three fathoms and turned in, leaving Doc up on watch.

16

Cape Cod Canal

In the dimly lit interior of the cabin, we probably would have slept all morning—but not with Curney on board. As the dawn broke, there was a terrific burst of oaths that came booming down through the forehatch to awaken us with a jolt. Of all the bad habits possessed by that man, no one could ever have accused him of being lazy or a sleepyhead. For years, he claimed, regardless of what hour he turned in, he was up at the first crack of dawn, which, obviously, was to be our schedule hereafter as long as Dave was on board.

Emerging on deck, we saw that the fog seemed to be thinned to the south, and there, not five hundred yards off our bow, was a glistening stretch of white sand looming up through the steaming mists, blending and disappearing into a wall of nothing a few points to each side. There was something about this, our first sight of land, that sent a pang of joy through every vein of our bodies.

Getting the compass bearing of an estimated perpendicular to the sector of the inner circle of Cape Cod's characteristic hook, we realized that we could get a rough estimate of our position by noting where on the chart a similar line would fit. We, apparently, were about twenty miles from the canal entrance.

A brisk south wind had picked up during the night, so we weighed anchor, hoisted sail, and soon were briskly scudding along on a port reach.

This was very pleasant. After a hearty breakfast of bacon and eggs, I picked out a comfortable spot on the bow of our ship and just enjoyed a few minutes of leisure. We were gliding through a quiet sea, well protected from the stiff south wind by the long arm of the Cape. Truman was running soundings, but our charts indicated clear passage and a soft sand bottom even if we did go aground. Streaks of heavy fog soon were settling in again, less and less frequently letting up for a few minutes so that we could see and confirm our estimated distance off the beach. Finally, we were back in the pea soup again, and from then on, it was the sounding lead alone to guide us.

Peering out into the wall of nothing ahead, I just sat there and tried to keep from dozing off. The fingers of vapor twisted and wound themselves into fanciful shapes, and I amused myself by imagining them as the strange forms that they vaguely represented. Suddenly, I sat up, startled, like a watchdog indistinctly hearing a foreign sound. Was I dreaming or awake—was that apparition or stark reality? Looming up in the clear water just ahead were the shapes of jagged, hungry rocks lurking insidiously only a foot below the peaceful water. Something shot through my body, and I almost screamed at Truman, "Rocks ahead!"

Truman threw the helm hard over. We all knew that the *Silver Heels*, with her full forefoot and large deadwood, was not particularly smart in coming about. Even though Truman's quick handling of the wheel had saved us from a crunching crash, a slower but no less certain disaster was in store for us. The wind, which had been for some time off our port quarter, would soon blow us onto what was now evident as a barely submerged jetty. The gentle ground swell would mercilessly raise and drop our tender bilges on those sharp fangs of rocks.

With not a second to lose, I picked up the anchor, threw it bodily overboard, and shortly felt the gratifying impact as it struck hard on the shallow bottom. Drawing up the line on a short scope, I made it fast and felt the hook bite as our stern swung around to it. Truman dropped the main peak, and we went aft to see where we had settled. There, within jumping distance, was the jagged jetty where, had all of this been timed one second later, our expedition would have ended. We were safe, but we knew that with the wind where it was, it would be impossible for us to sail away—a backslip of six feet would pile us on the rocks.

We could not see shore in the fog, but we sensed that we must be fairly close to it. Truman and Doc decided to row in on the theory that the jetty would probably lead to the beach and by following it in and out, they would not get lost. They soon disappeared into the heavy vapor, and we just waited. Dave was sitting on the cabin top, deep in thought.

An hour later, we heard approaching voices to shoreward and, shortly after, saw our dory looming out of the fog followed by a yell as Truman and Doc simultaneously spotted us.

They climbed on board and remained smugly silent until we all (except Dave, who had no conception of what was going on) were comfortably in our big cockpit. Then Truman, with a grin, issued his report: "We are practically at the mouth of the canal." Rechecking our chart, there was no designation of a jetty.

The chart was only fifteen years old.

Uncle Hal now had a surprise for us. In the cockpit, locked, he had tucked away a one-gallon jug of white gas, intended for the lanterns, but fully capable of feeding our engine. This would give us all that we needed to start up, round the jetty, and get us into the gas dock that Truman had learned was just inside the canal mouth.

In a few moments, we heard the familiar roar of the engine and then moved ahead and picked up our anchor. Cautiously, we followed the jetty to its outer end and then moved off through the fog, Doc on the bowsprit and Truman at the wheel, ready to throw her in reverse in a split second. We headed slightly seaward at the same time, hoping to pick up the canal entrance buoys, but probably groped around for fifteen minutes before we sighted one. This, from our charts, gave us an exact position, and we laid a course onto the canal mouth. We knew now that it was just a matter of minutes before our gas would run out, but there seemed to be no apparent reason why we should not chance making it with the thought that we might as well anchor wherever we ran out as here. That was the most foolish thing that we ever did.

None of us had ever been in the canal before, except Dave, of course—our supernatural navigator with his special extrasensory powers. He was still in deep communion with some strange spirits on the cabin top. The last thing we wanted to do now was to consult him.

I was at the wheel, and in a few minutes, the anticipated first cough of the engine came, followed by a brief pickup, a few seconds of sputtering, and then silence. Truman was forward preparing the anchors and their lines for quick unreeling. Doc started to take a sounding, and a moment later we heard him exclaim, "My God!"

He had been calling out two, three, and four fathoms for so long now that it was quite a shock to find the ten, twelve, and fifteen marks racing through his hands. A moment later, he called, "By the deep, sixteen."

We must have been in the deep ship channel. From the drag of his lead, he ventured that we were in moving water, but it was too deep to anchor here and, furthermore, this was the last place we wanted to park—without power in thick fog in the main Boston–New York ship channel. Now that we were in it, we might as well drift leisurely along until we either entered the canal or approached one of the banks of the shore. We could see nothing.

And then suddenly, through the blowing clouds of fog loomed up the canal entrance. We were racing toward it, swept helplessly on in a current with the ferocity of a millstream.

Cape Cod, as we later learned, is affected by a strange tidal phenomenon. With total disrespect for the moon, the tide on each side operates entirely independently and considerably out of phase with that on the other side. Although the sea dogs of that area invariably say that no one can explain why, as is the case with most tidal actions, it does seem apparent that the strings of islands below the Cape divert and redivert the ordinarily normal surges and resurges of the moon-attracted water with resultant variations that are far too complicated to explain mathematically. At any rate, the tide can be high on the north side of the cape and low at the south side simultaneously, or vice versa. As the canal has no locks, the result is that a swift current pours through it, alternately south and then north. The former was now hurtling us along at an alarming rate.

Down we went, recklessly passing the deep-cut sand banks and occasional shacks along the edge. We seemed to be drifting toward the left side of the channel and only hoped that some strange fate would swirl us over toward the edge so that we could find a moderate depth in which to catch our anchor. Truman and Doc had hoisted the stays'l, which was now drawing slightly, giving us a bit of steerage gradually edging us to the left bank.

Suddenly, looming up ahead was a sight that struck terror once more into us. There was the gas dock! Strung all along it, securely tied, were eight or ten fishing vessels. We now seemed to be plunging straight toward them.

"Better let 'em go, Truman," I called, and a moment later, overboard splashed our main anchor, dropping straight down like a plummet. In spite of our terrific speed in relation to the shore, we were practically motionless with respect to the water. Truman must have had fifteen fathoms of line on the anchor.

He called back quietly, "She doesn't bite." The anchor very possibly was not even touching bottom in the deeply dredged and constantly current-scoured canal. Surely and quickly, he prepared another coil of line to add to give it more scope.

Just at that moment, there was a bloody scream from the cabin top—the scream of a maniac. Dave leaped to his feet on the after edge of the companion hatch, released an oath that ran a shiver down the spine of every man on deck, poised like a tiger at bay, and then leaped at me.

He landed like a gorilla on my shoulders, and we both crumpled down and rolled into the corner of the cockpit. We each tried to regain our feet. Dave lunged at me again, striking with a haymaker that I diverted into my body but that almost broke my ribs. I jumped back to give the wheel a turn—there was no one there to help. Doc and Truman were frantically uncoiling the auxiliary anchor line. Uncle Hal was bending another line to a second anchor lashed to our port side stays.

Dave drew back and paused momentarily to advise me in no mild manner, "I'm going to kill you!" And then he dove at me with a second vicious wallop at my body that, gone home, could have pushed in my whole chest. Diverting this with my left arm, I followed through with a right to the side of his jaw, which would have terminated the tussle for any mortal man. It had my whole bodyweight behind it, plus the impact of his lunge. It didn't seem to faze Dave but almost crushed my fist (for a week, I was in agony with every move of my hand). I was afraid. I couldn't hit him any harder, nor could I again use my right hand.

Dave dove in close, and we went into a clinch. I managed to get his head under my arm with the thought that I might get some sort of stranglehold on him—and then I remembered Pōhaku.

Pōhaku, who had lived in the Hawaiian Islands, had as a boy been taught the art of jiujitsu. During our days together, he had taught me many of these holds and nerve spots, which I had never had occasion to use except in practice and play.

Using the knife edge of my hand, I gave Dave a short, sharp staccato-like blow on the back of his neck. He crumpled on the deck.

Suddenly, Truman's anchor caught—probably a rock on the bottom—and the vessel swung around and drew up with such a jerk that all of us were thrown off our feet. We had little time to spare—only a few lengths away was the first of the group of vessels that we would soon have smashed to matchwood.

Truman made fast another line to the holding anchor hawser, and we snubbed it on the bitts to ease ourselves downstream to the first fishing schooner. From here, we ran a line to the deck and soon were able to ease her along to the gas pump.

Dave was still out, and I put him on the dock. He was our key to the *John Dwight*, but what good would that $350,000 be if we were all killed in getting it?

He came to in a few minutes, and I told him that he could not again board the *Silver Heels*.

"Where in the hell are we?" he asked, aware for the first time that we were not still at sea and apparently now considerably sobered up. Dave was very apologetic, giving us our first clue that his hold on us was not too tight. He wanted to talk it over.

It turned out that he, in his alcoholic delirium, had imagined that I was going to cheat him out of his share of the *Dwight*, that I had all of the controls—the boat, equipment, and finances—and that once we found the wreck, we would simply forget Dave Curney. I didn't quite understand why he would be suspicious that we could find her without him since he had for so long bragged that he alone knew where she was. I assured him that we had no such intentions. Soon, he admitted that he had no reason whatsoever to suspect that we had any such evil plans. "I was just drunk, I guess," he admitted.

We shook hands and then all turned to make fast, at least temporarily, an extra tank that we had. We then gassed up with a now ample reserve for the balance of our journey. A few hours later, we again pushed out into the raging current, but this time with the engine throbbing assuredly and with full steerage to our helm. Our own speed, plus that of the current, sent us hurtling on our way, the canal banks racing madly by.

We were now getting well organized, were familiar with our vessel,

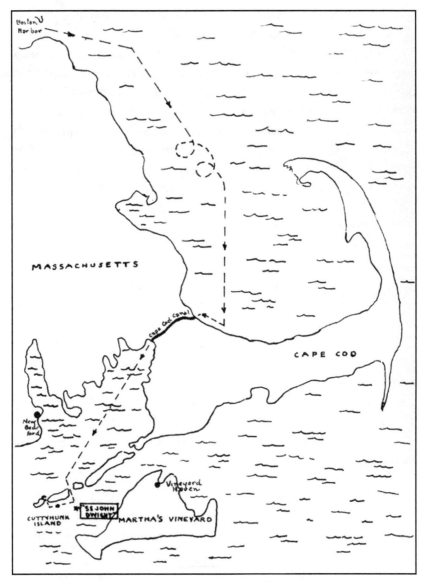

This map, hand drawn by Nohl, shows the path of the *Silver Heels* from Boston Harbor (upper left) to Cuttyhunk Harbor (lower left). The approximate location of the *John Dwight* wreck is marked with a star at lower left.

and had had enough close shaves to put our seamanship on a sound conservative basis. We chugged along, negotiating the pass with no trouble, rather offsetting our bad start. Our last channel, the narrow zigzag passage between Cuttyhunk and Nashawena Island, was just six inches deeper than our keel and just a few feet wider than our beam, but even with a strong current flowing, we felt our way through like veterans. This was a pass we were to navigate many times in the next few months. We triumphantly entered what was to be our base port—Cuttyhunk Harbor. It was here, on the Coast Guard station of this almost barren island, that Roland Snow, back in 1923, had seen the *John Dwight* disappear beneath the waves.

The following morning, we were to proceed out to the wreck—by noon, certainly, I could be in her cargo hold. It was hard to go to sleep that night, so close to such a substantial treasure.

By sunrise the following morning, we were on our way. We were wondering just exactly how Dave planned to find the wreck. He was practically blind, at least for anything farther away than the end of his nose. But during all of these months in which we had been preparing to sail, he had so dogmatically overridden my frequent attempts at questioning him that I had finally decided not to bring up the subject again for fear of further antagonizing his exaggerated distrust of our motives. Now, however, we were rapidly approaching the area of probability, so Dave and I went into a huddle.

"You know," he started out, "this old man can't see quite as well as he used to."

I knew.

"Gimme a piece of paper," he asked.

I found some, and he took a stub of a pencil out of his pocket and started making some queer little marks on it, almost touching his nose to the paper as he drew.

"On Nashawena Island, you'll see a V shape in the bluff. You line the point of the V up with a shack on the island and you have one bearing!" Dave was now sweet as sugar.

"On Cuttyhunk, you line up the big rock on the beach with a clump of trees—you'll see them—and where those two cross, there's our wreck!" He made a lot of funny little scribbles on the paper, which added nothing to the clarity of the situation whatsoever. We knew the shape of a V.

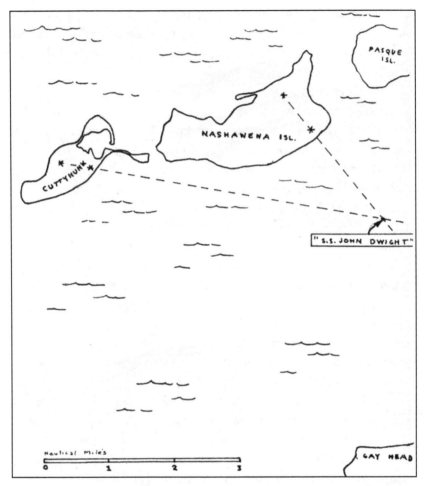

According to Dave Curney, the *John Dwight* wreck was located at the intersection of sight lines extending from Cuttyhunk and Nashawena Islands.

We strained our eyes. We brought out the marine glasses and scoured the landscape in detail. We could not discern a single one of the landmarks. It was a beautiful clear day—never would the visibility be better. After a few hours of searching, moving in and out from the shores, thoroughly discouraged, we headed back to Cuttyhunk.

With the *Silver Heels* secure to the dock, we spent the rest of the day calling. We talked to innumerable fishermen, natives, Roland Snow, and the entire gang at the Coast Guard station. Everyone was friendly, and we felt that we were going to enjoy Cuttyhunk. Here, truly, was an island of the sea, and everywhere, permeating the remotest corner of its rugged terrain and every living thing thereon, was the atmosphere of salt water and ships.

We learned two things.

First, not one of Dave's bearings would be of any value to us. The shack on Nashawena had long since been torn down. The V, if it ever represented much more than a temporary discoloration in the island's foliage, would by this time be grown over, eroded down, or blended into its background a thousand times over in the long years since 1923. There had been timber cutting on Cuttyhunk, and the clump of trees of 1923 was now as ephemeral as yesterday's clouds. The rock—yes, there had been a big rock, but that had been constantly moving in big storms, and it was just one of many on this rocky island. Out of the four points, we had none.

Secondly, we learned that there was probably no one on Cuttyhunk who didn't know considerably more about the position of the *John Dwight* than Dave. Everyone was interested. They each had their pet theories as to where she lay. Snow had seen her go down and could give the general direction from the Coast Guard station with an estimate that she was six or seven miles offshore from there.

Dave had to get back to Boston. We now had a long list of supplies, materials, and equipment that we needed, and it seemed advisable that we put in to Vineyard Haven, which was Dave's home, for repairs, overhaul, and redesign of a good many details on the *Silver Heels* before starting the apparently serious job of searching for the *Dwight*. I would go to Boston with Dave, buy our supplies, and take the steamer back.

The next night in Vineyard Haven Harbor, he seemed to us to be in better spirits than we had seen for some time. He invited us over to his house for dinner, which we were soon to find was no minor treat.

Truman and Doc had some other plans, and so Uncle Hal and I accepted. Dave's wife, Hat, a woman of considerably more avoirdupois and height than her husband, proved to be as sweet and kind a wife as ever lived. More than that to us at the moment, she was supreme as a cook. She served us a dinner that would be the envy of kings, a meal that we enjoyed as much as any we had ever had. And we particularly enjoyed it after the comparatively crude turnouts from the *Silver Heels* galley for the past months. Dave was friendly and kidded all of us and himself in front of his family.

After we finished our dessert, he got up from the table and went over to the front windows, opened them, closed the blinds, and then closed and locked the windows like a fidgety old maid locking up for the night. In turn, he did the same to the other windows and then, with the same cool premeditated manner, locked the front door before our puzzled eyes and turned to watch me while he put the key in his pocket.

On Hat's face was stark terror. Dave sauntered over to the table, picked up a huge bowl of beans that we had only been able to sample in the bountiful meal, held them high over his head, and poised them there momentarily before his silence broke.

"You #?&#?¢@&, you, I'm going to kill you right now!"

The bowl was hurled to the floor with a splash of china chips, tomato sauce, pieces of pork fat, and sticky brown beans that sprayed over the entire room. The ferocity of that blow apparently was an indication of what he was now going to do to me.

Hat was screaming hysterically. "Dave—Dave—Dave—please don't! Dave—please!"

Dave determinedly walked out into the kitchen and, in a moment, reappeared with a huge butcher knife. The hospitality of his dinner table apparently was now over.

I had no defense whatsoever against that blade of almost swordlike proportions and, furthermore, had no desire to fight with an intoxicated man.

I ran upstairs, the only place to go, and he followed. I came down the back stairs, and the chase was on, I being careful not to get in any corner or room without an additional exit. It was not difficult to elude him since he was sufficiently inebriated to be very clumsy.

Down in the kitchen, I found the back door simply hooked and walked out. Hal followed, and we walked leisurely back to the dock, a stream of oaths following us as Dave gave up at the back door.

The *Silver Heels* was anchored a half a mile off the dock, and we felt quite confident that Dave would end up sleeping on the davenport rather than pursuing us further.

17

WRECK BELOW

The next morning at day break, we were wakened by the customary commotion up on the deck. We heard Dave's voice in the cockpit as he instructed the man who had rowed him out not to wait for him. He apparently gave the man some money for services performed, which the oarsman thought was too much. We had to admit that there was nothing stingy about Dave.

There was no cause for alarm, for he was perfectly sober and in good spirits as he called out his typical morning salutation, "All you guys do is sleep!" After all, it was almost five o'clock.

Dave, sober, was as friendly and likable as anyone could be. He came below and started talking his head off, making a gesture at an apology for last night. "Guess I was a little drunk last night, huh? Don't pay any attention to me when I'm drunk."

I was planning to go to Boston for supplies, and Dave had to get back to his job up there. We had planned yesterday to go up together, and I saw no reason why we shouldn't even though he had tried to kill me last night—but from now on, it would only be the sober man with whom we would have any relations whatsoever. Feelings or no feelings, his name was still on the *John Dwight* contract.

I noticed that he was limping quite badly. By the time we reached Boston, his ankle was alarmingly swollen and he showed visible signs of pain, the first trace of discomfort I had ever seen in him. I wondered if last night he had fallen out of the window or down the stairs without knowing it. Thinking it might be sprained or broken, he decided it would be best to have the doctor look at it. On the other hand, he was concerned about his job, but he did not feel capable of doing it in this condition.

Dave's job was in the water under a large building owned by the United States Army in South Boston that served largely as a warehouse, with facilities for direct loading and unloading for ships. The entire structure was built over the water on tremendous piles.

These supports were spaced only a few feet apart in even rows with several feet between the rows. The building was, in base area, the size of a city block, thus presenting a veritable forest of many thousands of long creosoted wooden piles.

The constant surging tides running under the building and over the muddy sand bottom stirred up a sediment that kept the water as black as ink. Whatever trace of light might otherwise try to creep in was completely shut off by the massive building overhead. Thus, a diver far underneath had to plod through the ooze by pure feel.

Periodically, someone would have to inspect every single one of those thousands of piles. The worm action and other deteriorating effects of the salt water were constantly eroding away their exterior. Originally, they were twenty-four inches in diameter. When that dimension had shrunk to eighteen inches, a pile would have to be replaced. Thus, the diver delving hundreds of feet under the building must, row by row, measure each one and report by telephone its number and diameter. There was a measuring caliper with deep grooves in it so that the diver could read the instrument by feel.

Dave was tops in a job like this. His handicap in vision was an asset rather than an impediment to him, for he had long since learned to handle himself and work with tools by feel alone. The pay was standard diving rate, fifty dollars per day, and the contract would go on for months.

I didn't blame him for not wanting to lose the job contract. In the professional diving business, it was a job here and a job there—most of them simply one dip. This was steady work—day after day—with equipment all set up, shallow water, and with full pay. It was a job that any diver would fight for.

It would be tough working with that ankle, and so Dave decided that if I would pinch-hit for him for a few days, he would go to the doctor and get off his feet until it was healed. The boys would be busy finishing the fitting out of the boat for a few days anyway, and we could use the extra money.

I reported to the foreman on the job.

"So, you are substituting for Dave Curney?" he repeated with a laugh. "Curney was fired yesterday—we don't want to see that guy's face around here again!"

I told Dave. I rushed through the shopping in Boston in hopes that I could get back to Vineyard Haven and we could shove off before the man of iron showed up. We didn't need him anymore, and although I intended to see that he got his just share of the prospective proceeds, I hoped that in the future we could elude him.

I loaded the new equipment in Doc's car and drove down to Wood's Hole and took the steamer from there. It was late at night when I arrived at the dock at Vineyard Haven, and I yelled my lungs out trying to stir someone on the *Silver Heels*, anchored a half mile out. There was no sign of any life anyplace, so finally in desperation, rather than sleep on the dock all night, I jumped into a strange dory, the only boat available not under lock and key. I had planned to row out, pick up our own skiff, and then immediately pull back to shore with it. I could then row out to the *Silver Heels* in our own boat. However, all except Doc, who was ashore on a "heavy date," wanted to know what luck I had had, and so I sat there in the cockpit for about ten minutes with the stolen dory's painter in my hand.

I was just getting ready to shove off when out of the night appeared another similar dory rowing frantically toward us with a very irate man, to put it mildly, at the oars.

He unleashed a storm of curses at me for stealing his craft. He was furious. Apparently, he had seen me take his boat and had gotten out of bed to pursue. I felt rather badly about this, since we were anxious to establish and maintain friendly goodwill with these people with whom we would be in contact for some time. I apologized profusely. He rowed off, still muttering, with his previous dory in tow. We then sat down in the cockpit and talked over our situation and plans and the problem of Dave.

A half hour later, out of the black night, we heard a boat rowing toward us. As it approached, we saw that it was Doc.

He had "borrowed" (without permission) the same precious dory! Soon after followed the same, very irate man.

We lay in another day and on the next hoisted the hook and headed back to Cuttyhunk to start the serious business of locating our treasure ship. Arriving back at the island harbor as the sun was about to set, it was with a glow of gladness that we realized we could now go about our work unhampered by Dave.

That evening, we went back to our best resources of information on the wreck and burrowed into them in full detail. We finally decided on an area of approximately one square mile as the place of highest probability for the find. We laid this out on our charts and assigned numbered buoys to each corner of the square with intervening smaller buoys along the four sides.

In the morning, we chugged out to find our square mile of water and started the process of planting the real buoys in the corresponding locations of the little Xs on the chart. As we lined up our landmarks on the shore, overboard they went, one by one, until we felt we had staked out our claim on Davy Jones's wealth. It was a real thrill to look back and see the lines of red and yellow cans stretching out and realize that probably someplace within this enclosure was a fabulous fortune that would be ours for the finding and retrieving.

The next job was to be no snap at its best—to thoroughly comb every inch of that area with our drags.

We worked out the plans and built a set of drags, so that with a single boat we could sweep a substantial swath of the sea floor. In shape, the drag was a huge isosceles inclined triangle with its apex at the stern of the boat and its base dragging over the bottom. Looking aft from the boat with it in action, we saw just two lines disappearing down into the water off each quarter, spanning an angle of perhaps thirty degrees. These reached down to the base corners, which were four hundred feet away on the ropes. The base of the triangle, or the width of our swath, was two hundred feet. Our square mile would average in depth from fifteen to twenty fathoms (90 to 120 feet).

At the two base corners were our spreaders, which would serve to keep the base rope stretched out and also keep the drag on the bottom. These were flat wooden planes, three by five feet in area, well weighted with lead along their lower edge. They were towed by a chain harness fastened to each corner of the rectangles in such a way that by moving our shackles from link to link we could steer the drag in any degree we chose for more or less sidespread or more or less diving tendency.

They really worked.

We hadn't moved fifty feet when they caught up sharply on some snag at the bottom. Overboard went the buoy and, in the minimum number of minutes that it takes to get the balance of the suit on, overboard I went, too.

Down I went, sinking like a plummet, but with every confidence as I felt that air blasting in from the compressor to quickly meet the rapid pressure increase. A minute later, I felt a soft sand bottom come up and strike my lead shoes. Transition in depth is difficult to judge except by indirect observations such as pressure increases, falling off of the light, and startling changes in temperature on bare hands; otherwise, there is no sense of relative motion in the wastes of blue-green space between bottom and surface.

A jerk on the line let the boys know that I was on the bottom and wanted all slack. Off I went, sensing my direction from the set of the tide current, which was flooding and running toward the sound.

A few fish gazed in goggly-eyed wonder at this strange invader of their world. Scattered clumps of weeds were swaying to the current. Dimming off into an indiscernible nothing in every direction was the beautiful blue-green water, visibly moving and sweeping inshore countless particles of seaweed and minute organisms.

Bending into the current, I lumbered along until I saw swooping out of the dim distance a bright yellow manila rope, almost incandescent against the dimly colored background. Following this along in a few more steps, I would find the objective of this dive.

There, looming up, was a low massive structure around which, I could now see, was entwined our drag.

It was a rock. A huge slime-covered rock.

Our next dragging operation took us only slightly farther. In about one hundred feet, the lines caught up sharply again. With hopes still high, down I went again, and there it was—another rock.

We started to wonder—how many rocks would there be in a square mile of ocean floor at this rate?

These rocks had not been huge structures towering high above the bottom. Perhaps if we set our base line a bit higher up on the side lines of the drags it would not actually scrape over the bottom itself and catch every tiny protrusion but just the larger ones. After all, that steamship would present quite a substantial mass above the sea floor.

This theory worked beautifully, and we found that we could set our snare for almost any size snag that we chose. However, we knew we must not be too eager, since the sought-for vessel might well be settled down into the mud or broken up and flattened out considerably after all of these years.

The next day, we saw a fishing boat approaching us. As it chugged closer, bow headed due on us, we all seemed to sense something was coming that was not good tidings. Fifty yards away, a din broke out above the roar of our engine, and we knew that we had been right. It was Dave.

He was spouting away furiously. Why had we sailed without him? Why hadn't we left a message with Hat?

No one paid any attention to him. He couldn't board the *Silver Heels* until the end of this sweep for fear of tangling up the propeller of the fishing boat on which he had taxied out with our submerged lines.

As we pulled up to the edge of our square and took our weigh off to turn for the next sweep, the fisherman came alongside, and Dave boarded us. I explained to him that we were going ahead with the job in the manner which seemed to us most suitable and that we would tolerate no interference from him. He would be welcome to watch; his suggestions would be heard; we expected him to help with the diving when we located her; and the contracts were valid—but after he had been drinking or at any time that he imagined himself as in charge of our expedition, he would not be welcome. Dave, now perfectly sober and also shocked by the loss of his government job, was very meek and receptive. Under these conditions, he was as likable a fellow as we had ever known. Fundamentally, the guy was made of gold—his only bad habit was that occasionally he would try to kill me.

Days stretched out into weeks. We combed and then cross-combed our buoyed square, but we looked out on it with only dismal hope, mentally imagining that we might keep expanding this to embrace the entire earth's ocean floor, some 150,000,000 square miles. Not on this diet!

Our fare was not that to please an epicure. We hadn't tasted fresh meat in a month. Practically all of our meals were at sea—we put into Cuttyhunk Harbor late each night, and we were underway early in the morning before

Dave Curney (second from right) dispensing detailed advice to Nohl, (center). From left: Truman and Uncle Hal, and to the extreme right, Dave's son, Donald.

we were fully awake. Doc barely had the morning stove hot as our bows rose and plunged to meet the first swells of open sea. Breakfast was served as we chugged along to our first buoy and, like all of our meals, consisted of only nonperishable foods—dried cereals and powdered milk. Lunch at noon was a cold sandwich when and if you had time to grab it, with a strict understanding that it must not slow up or interfere in any way with the work at hand.

Our supply of meat consisted of case after case of canned corned beef, which, I understand, can be cooked up into quite palatable dishes. With Doc, however, and his limited culinary experience, it was just a matter of opening the can, laying out a pile of unbuttered stale bread, and calling, "Here it is!"

To vary the corned beef regimen, we had, almost nightly, what we had once vaguely remembered was an epicurean treat. Cuttyhunk was a large lobster fishing port, and all around us, as the sun was setting, in would come boat after boat laden down with the creatures still dripping the cold water of their native depths. We had done many a favor for these fishermen, and they, in turn, would constantly bring us bushel-basket loads of the delicately flavored crustaceans.

The first night that this happened, we enjoyed them with as much or more delight than would the most fanatic lobster fan. They were particularly pleasing in the atmosphere of their native habitat, and we certainly could have no complaint about their not being fresh.

I had had a very limited number of lobster dinners in my life and even on these occasions only because someone else had selected (and had been paying for) the meal. It had always seemed that for the price, a lot more food value could be obtained in a restaurant by ordering a sirloin steak than those over-expensive tidbits. Now, suddenly, we had the crustaceans in any quantity that we could eat them, and we soon would have been willing to trade a hundred dollars' worth of lobster for one thin slice of roast beef.

The thought that we were stowing away ten dollars' worth of meat apiece with each supper soon was of dubious consolation. We ate only the tails since the claws involved extra labor in extraction. Finally, we refused even to accept them as a present, and we were glad to go back to our corned beef.

Everyone was growing restless. The life on Cuttyhunk, although rich with the lore of New England and the sea, was growing tedious to our gang. Our cash reserves were rapidly thinning with the many unpredictable exigencies and the daily expense of feeding our hungry mouths and satisfying the insatiable thirst of the engine. We were enlarging our square-mile area of dragging but, as we did so, were noting that just one two-hundred-foot swath around it was a trip of over four miles.

The definite decision to quit had not quite been reached, but it was clearly apparent that we were going about our operations from day to day with the knowledge that each one, as we headed back to our nightly base, might be the last. Realizing this, I was pushing the boys even harder, getting out a little earlier and staying until the last possible minute.

We finished the perimeter of our enlarged square late on one of those discouraging afternoons and, by a look at the low-hanging sun, knew that it would be too late to start another side with the rapidly dimming light. We decided to head back to Cuttyhunk a little early, but since we had extra time, we left the drag on the bottom, merely sweeping a chance course as we laboriously chugged toward the mouth of the pass.

Suddenly, the drags caught sharply up, apparently on some massive structure, and the *Silver Heels* shuddered to every timber in her massive frame. The engine groaned as we came to a dead stop and then labored desperately as it churned up a stream of white water aft from our stationary boat.

Quickly, we hopped into our dory, pulled aft over the theoretical position of the snag, and swung our lead overboard. Truman was singing out the results of successive soundings as Doc skulled the boat about.

"By the deep, nineteen (114 feet)."

"And a half, nineteen (117 feet)."

"And a half, nineteen."

"By the deep nineteen."

"And a half, eighteen (111 feet)."

"By the deep, nineteen."

"By the mark, twenty (120 feet)."

"And a half, nineteen."

And then came the call that sent a pang of joy through all of us: "By the mark, fifteen (90 feet)."

This was suddenly four fathoms (twenty-four feet) shallower than the area which he had been sounding with only a horizontal movement of a few feet. This meant that there was a vertical rise of that distance and could mean only a submarine cliff or a wreck. The former was highly improbable, since the constant motion of the tides and current, particularly here, tends to level everything, sanding and filling depressions and eroding down ridges.

I asked Truman to try the tallow. Inserting this in the cup-shaped depression at the base of the lead would pick up crumbs of the bottom, and we could, by their study, determine much as to what we would subsequently see down there.

"By the deep, nineteen!" sang out Truman again as he dropped his lead off the imagined wreck. A moment later, pulling up the dripping line hand over hand, he grasped the lead, studied it for a moment, and added, "Mud bottom!"

Doc skulled him back to the area of the lesser sounding and over went the lead again. Truman bounced it energetically up and down a dozen times to penetrate what would be a surface scum if this were our prize, and checked his depth designation: "By the mark, fifteen!" This was on the wreck.

Up came the lead, and after a second's study, Truman turned and grinned, "Flecks of white paint!"

Unless God was painting his rocks white in these waters, here was at last the elusive end of the rainbow.

18

End of the Rainbow

The pass into Cuttyhunk Harbor was strictly for daylight navigation. On this barren outpost, no lights had been provided for later-returning mariners. Following an exact channel through which was running a heavy tide rip, we had just six inches of water to spare under our keel, and we always breathed a sigh of relief each time we negotiated it successfully. To date, we had not so much as scraped the bottom at any point.

The light was failing rapidly, and it would have to be a fast run to make it before the last gleam disappeared from the foreboding sky.

"To hell with going in!" we decided. The seas were battering our hull mercilessly. A heavy ground swell was running. The currents, in full spring tide, were racing by like a millstream. The angry black clouds warned us of nothing but trouble.

All of our heavy weather buoys were well planted on our estimated probable area. We were quite a distance from there now, apparently on the wreck, but we had nothing very substantial to mark this prize spot. Our tiny improvised cans would do it temporarily, but we were afraid they would be carried away in the rigors of the forthcoming night. A low blowing mist had almost shut us in so that it would be impossible to take shore bearings. We didn't know just exactly where we were because our course and speed had been indeterminate since we left the last buoy with the retarding drags out.

The only buoy on which we could really count was the *Silver Heels* herself. It would be a nasty night ahead, but we would be in position at daybreak and could get our anchors laid during the evening—and how nice it would be to sleep within 112 feet of a half a million dollars.

All of this time, Uncle Hal had been idling his engine to keep the snagged drag taut and insure against its slipping off. In the dory, we now carried our anchors out, one up into the racing tidal stream, and another down current (which would be up when the tide reversed in another six hours). Two more were placed as steadiers off our starboard bow and the other off of our port quarter to form a thin X, which would meet any possible current or wind changes that might occur.

With all hooks biting, we warped the *Silver Heels* on her windlass back and directly over the position of the wreck and made fast.

As a final check, Truman dropped the lead again, this time over our own bow instead of from the tender, and sang out, "By the mark, fifteen."

We were secure and over our long-sought treasure.

Next, overboard went the descending line in the same place, a fifty-pound hunk of iron on the end of a one-inch-diameter rope. Down this, I would slide in the morning to land directly on her deck.

That was a ghastly night, offset only by the joys of many fanciful dreams as to how we would use that money. After dark, the wind grew to almost gale proportions, sweeping in from the open Atlantic to kick up seas that tossed our craft like a tiny cork.

She would dive almost from under our breathless bodies to meet each oncoming gaping trough and then shudder to her very core as her bow dipped deep into the following wave. Up she would leap, pressing us hard into our poorly padded bunks, to be stifled by a violent jerk as her never-relenting anchor line drew taught. Combined with this was a violent yaw and roll that twisted us around in a figure eight of constantly varying pattern. Every timber in her hull seemed to scream for mercy as the giant seas strove to test her for worth, seemingly avenging our forthcoming attempt to defy the sanctity of her depths. Every frame chimed in with its peculiar creak and groan, to join the symphony of punishment. The violent contortion had apparently loosed several seams, for there were scores of gallons of water in the bilge sloshing around angrily. Every dish, every tool, every movable thing on board was banging restlessly against the confinements of its enclosure.

We were all half sick and almost exhausted by morning. Rather than a night's rest, it had been more than a day's work just to stay in bed.

At the first trace of daylight, we were all up and on deck, overjoyed to see that not one of the many uncontrollable things that might have happened did happen. All anchors had held without giving a foot. The descending line was still on the wreck. At last, God was on our side.

The boys started getting me into the gear while Doc finished up the breakfast. With breastplate and dress on, I swallowed my cereal, dried milk, and cocoa, and in a few minutes was ready to leap overboard.

After slipping on the forty-pound lead-soled shoes, I climbed up out of the cockpit and over to the diving ladder. This was a ticklish operation with the boat rolling and pitching as violently as it was, and accordingly, Truman took extra precautions to secure my lifeline to one of the quarterdeck cleats. It would be very easy to lose balance with the heavy gear and fall overboard, which, without helmet but well weighted down with lead, might be a rather embarrassing experience if my line were not yet tied to me.

The helmet went on with window still open. I felt the boat tremble as the massive engine started up, and in a few seconds, as Hal opened up the valve, blasting into the interior was the familiar roar of air. The last adjustments were made, and the window was closed. I heard two slaps on the helmet top, the signal for "All is ready." I started down the ladder and, half submerged, was almost swept off it by an angry breaking sea.

I, like everyone on deck, felt a bit woozy from the violent night. Although no fish had been fed to my knowledge, we all felt lightheaded, irritable, and exhausted. Riding at anchor is always much harder to endure than running in the same sea. Thus, it was almost a revelation to suddenly find myself in a world of peace and quiet. As a reminder of the turmoil above, the descending line in my hand was alternatingly straining taut and slacking off with the rise and fall of the *Silver Heels*, but I myself was free of all violent movement, simply sinking silently down into the darkening depths.

All of my lines were slack, and I was falling like a plummet. I could feel a small wind blowing through my Eustachian tubes to equalize the pressure in my inner ear. The water was darkening so fast that the midwater vista before my eyes was dimming almost like a motion picture fade-out through the iris-shaped front window.

Suddenly, something soft came up to hit my feet. I regained my balance and, leaning into the tide, noticed with disappointment that I was not on a

Nohl dressing for the *Dwight*.

Nohl entering the water from the *Silver Heels*.

wreck—below me was an oozy mud bottom. I backed off a step and noticed that there was our iron fifty-pound weight about three quarters buried in the mud. But, before I could do anything, I must send up my signal.

Groping for the control valve, fastened to the left stud of my breastplate with a short lanyard, I turned off the incoming air for a moment above the din of which God, Himself, could not think.

"On the bottom—OK—on—the—bottom—OK—on—the—bottom— OK," I called up over the phones, which under these pressure and reverberation conditions could never produce what might be called a high-fidelity signal.

"On—the—bottom—OK," repeated Truman in confirmation and added, "On—your—lifeline—nineteen—fathoms—nineteen fathoms— nineteen—descent time—fifty-two—seconds—fifty-two—pressure— fifty-three pounds—fifty-three."

"Fifty-two—seconds—fifty-three—pounds," I confirmed and opened the air valve for a terrific blast of air.

Apparently, the descending line weight had slid off of the wreck or had been jerked off during the night by the movement of the boat. However, the wreck (or whatever it was) must be nearby since the *Silver Heels* certainly had not drifted. Not having the slightest clue in which direction to start other than with the iron weight as the center of probability, I started circling on the bight of the descending line. Fathoming off two lengths of my outstretched arms, I plodded around the circumference of the corresponding twenty-four-foot diameter circle with no observation other than seeing a lot more water. The weight was over twelve feet from the wreck—the only conclusion that I could draw thus far.

Next time, I decided to double the diameter, going out on a twenty-four-foot radius for a forty-eight-foot-diameter circle. The wreck was big enough that I would not have to cover every square inch of the bottom to find it.

I trudged painstakingly along, wearying of the rather dull scenery. An occasional fish peered into my window and then suddenly scurried away as he sensed potential danger in the strange-looking rubber- and copper-bedecked organism that was a stranger to his world. The tide was running somewhat, and its sweep was picking up enough mud to make the visibility poor. The light was dim, and it was with difficulty that I could see the bottom from an erect position. However, in the more natural inclined walking position, leaning forty-five degrees forward, I could see the waving weeds and marine flora beneath my helmet as I passed.

Suddenly, I sensed the presence of a huge hulk ahead. It is difficult to describe this subconscious apprehension—but some instinct seemed to tell me that just before me was a mighty mass of something. I paused, peering into the blue-black wall of void before me, but could see nothing. I took another step and understood.

Apparently, the light was fading, giving me some intuitive sense of approaching an overshadowing form. Another step and an obvious degree of darkening was discernible. Something was shooting through me in the joy of expectation. It was now very black. I noticed that the water was hardly stirring. I must be almost under the very shadow of the hull.

One more step, and there she was!

Looming up before me was a solid sheer wall scarcely visible under its heavy coating of green marine slime. A step closer, and I could touch it.

Stretching my arms out as far as they could go, I pressed close to it and felt her shape. It was not a vertical flat wall—I could feel the gentle turn of her bilge. Reaching fore and aft, I could barely and yet definitely discern the gentle sheer of her run. It was a ship!

My next duty was to establish this point for temporary future reference. Turning, stepping away from the hull, and leaning backward, I dug my heels into the ooze and tugged on the bight of the line to drag the weight over. As it plowed through the soft mud, I felt it hop, skip, and jump toward me. This was all a lot of wasted energy, for I should have landed on deck in the first dive, but now I had better continue to approach from here. Not knowing what a tangle her deck might be, this was actually the safest way to do it.

I now shut off the air and called to Truman: "Approaching—wreck—approaching—wreck."

Truman came right back in confirmation: "Approaching—wreck—all—OK—all—OK."

"All OK," I confirmed and opened up the control valve admitting another blast of accumulated air, which entered the helmet with a roar. Immediately afterward, I started closing down my exhaust valve to build up buoyancy. By taking the full output of the compressor with the intake valve open and by almost cutting off the constant surge of outgoing air, as now was evident by the decided decrease of gurgling bubbles wending their way back to the surface, I could cut my weight.

This is an almost eerie sensation to a gravity-accustomed animal. Gradually, the diver feels his body righting itself to the buoyant air accumulating in the upper dress about his shoulders. Soon, I was floating almost vertically with no power to bend from that position. My weight on my feet was falling rapidly off until the familiar moment when I was not sure whether I was standing on the bottom or just hanging up in space like a ghost. I glanced at the hull and saw that I was now slowly moving upward. Like taking off in an airplane from a smooth runway, it was impossible to determine the exact moment when I left the bottom other than by observation of some relative movement, as the plane passenger would, with a lump in his throat, notice the ground dropping from beneath him. Faster and faster the slimy hull raced by my window, my small area of vision making me feel very much like a fly crawling up over the side of an elephant's back.

I was in no danger of missing her gunwale and shooting up into the open water above, since I had the descending line securely in the crook of my arm and was sliding along it like a trolley. At any moment that I wished to arrest my motion, a grab of this would stop me—at least to the extent of the fifty-pound weight, which was far in excess of my present buoyancy. However, to avoid exposing too much of the bulk of my suit to the strong tides running over the wreck, I took a knock at the exhaust valve button on the helmet interior. This is a convenient extension of the valve stem, making it possible for the diver to operate it without using either of his over-busy hands. A blob of air bellowed out, and its loss was almost like touching the brakes on a speeding automobile.

Everything was well under control. In a moment, I saw light water and then the ship's sheer boards and waterway stripping. As my helmet emerged above the deck, I felt the flow of the tide. I grabbed her rail. I noticed, a few feet to my right, an aperture which, I had remembered from her pictures, was probably cut for swinging cargo inboard, and so I was undoubtedly immediately opposite her main hatch. Working my way along, I soon was adjacent to the opening and stepped through for my first formal appearance on deck.

My first job was to make certain that I could find this spot again. I turned outboard, leaned over the rail, and hauled up the fifty-pound hunk of iron, much as sailors would haul in the lead line. I deposited this in the gutterway and then, taking in some slack, made two half hitches on the bight to the rail. Whatever might happen from now on, I could slide down this line and reach this same spot on a subsequent dive. My only question now was where on the vessel this spot might be. Going out on the bight again, I worked my way along the rail, not yet knowing whether I was on the port or starboard side or walking forward or aft.

Because of the poor visibility, I decided to crawl on my hands and knees, since I could easily step and fall into any deck aperture before seeing it.

It seemed as if I had crawled fifty feet before any object that I could identify showed up. I ran into her fo'c's'le head, the raised forward deck, and then knew that I had been on her port side and was walking forward. While up this far, I stepped up on it and saw a clutter of anchor chain, the flukes of a huge anchor, a pile of wreckage, and her fairleads and bitts. At some later date, I would explore her fo'c's'le, but not today—first, I wanted to find the cargo hatch.

I crawled aft along the now familiar port rail, being careful to occasionally stop and check my air hose and lifeline for fouling, since her mast and rigging might still be formidable hazards in the blind area above. Back at the opening in the rail, I imagined that the hatch would be immediately adjacent. Starting out athwartship, I soon was pleased to find that my calculations had been right.

Before my window was the hatch coaming.

I looked over it into the large open void—the hatch covers had apparently floated away—and peered down into the inky black waters of the cargo hold. What could be down there? Was this the end of the rainbow?

19

The Atlantic Leaks

I paused to send my periodic signal to Truman, a required safety precaution that we always observed on our boats. "At—fore—hatch—coaming—at fore—hatch—coaming," I called.

Truman confirmed my call and then added, "Fifty—minutes—prepare—to—come—up—prepare—to—come—up—fifty—minutes—wind—picking—up—wind—picking—up."

It was hard to believe that I had been down for that length of time, but I had taken quite a walk. I had a long decompression ahead of me—almost an hour to get up—and every extra minute that I chanced down here would require an extra one in addition on the trip back, so, I guessed, I had better get started. On the next dive, I could be at this same spot three minutes after entering the water.

However, Truman had said fifty minutes. I could stretch this to an hour if, at the end of that time, I was ready to leave the deck with all lines clear. I could use another five minutes. There wouldn't be time to attempt the descent into the hatch, but I could explore her deck a bit more. Before I dropped down into the hold, I should know where her mast and rigging were, anyway, and this might be a good time to look around. I had seen sections of wire rope on several occasions, indicating that mast and stays were in a jumble all over the deck—and these would be things in which I definitely did not want to tangle. Crawling along the hatch coaming, I worked forward to make sure that I had made no mistake on this identification. With visibility not over three feet, was it even possible that I was on her starboard side and looking over into the deep black water of the wreck?

These suspicions were soon allayed as I came to the right-angle turn of the forward port corner of the coaming. Around this corner and off to the starboard, the deck continued, and so my assumptions were not incorrect.

I decided next to look for her mast, which was midships just after of the cargo hatch. I moved toward the stern again, along her coaming, until I came to the corresponding after corner and then struck out on a dead reckoning course for where I thought the mast might be. I was crawling along as fast as I could, realizing that the last few minutes of my allotted time were rapidly slipping away and that I must get all clear and back to my base (where I had half hitched the descending line to the rail) before the hour had passed.

The deck was covered with debris—cable, broken pieces of decking, rotten rope, and all sorts of miscellaneous junk. At the moment, I had no time to examine this material—probably settled there as fragments of the explosion—but later might find many interesting clues here. Now, my sole purpose was to determine whether the mast was still standing and, if not, where it was.

Crawling over the debris, I was constantly catching my chafing pants (the outer loose canvas pants worn over the dress to protect the former from tearing) in small splinters of wood or cable, but I had little fear that there could be any damage to the dress itself.

My feet would occasionally catch, too, but since it was impossible to see them that far back in the murky water, the best thing to do was just clumsily shamble along and drag myself through the debris like the proverbial bull in a china shop. Because of the coarse construction of the diving dress, chafing pants, heavy leather shoes, and many other encumbrances, it was difficult to tell just what was dragging where.

Something seemed to be snagged again, apparently on my left leg or foot. A gentle jerk didn't seem to dislodge it, and so I gave a more vigorous one. In too much of a hurry now, instead of backtracking, I gave quite a violent pull—and then experienced a terrifying sensation.

Water—ice-cold water—was pouring, in gallon quantities, into the dress. I had torn a gaping rip in my left leg.

I sat upright quickly to prevent the frigid fluid from flowing into my upper dress and helmet. I had been not too warm before, with my bare hands exposed for almost an hour and with one less suit of underwear than I could have worn. But now I was shaking as soon as the water hit my body.

However, there didn't seem to be anything really serious to worry about except for an hour of misery ahead—it would take that long to get up, and an hour in ice water would be preferable to a possible case of the bends anytime. I closed my exhaust valve to build up a bit more pressure in the dress and to force the water level down. I stood up to keep my helmet as high as possible and started to trudge back to the rail.

The water was now up to my armpits and sloshing around as I walked. Although the surface (over one hundred feet above my head) was warm enough for a refreshing swim, this felt as though it were about the temperature of melting ice. The level could not rise any higher, as in an open helmet, for the trapped bubble of compressed air in the headpiece forced it down.

I was shaking violently with the cold and decided to chance going up from exactly where I was. The walking was tough—my time must have been up on the bottom—and I couldn't stand that cold another minute (although I knew I would for sixty more and like it).

I shut off my air and called Truman. "Haul—me—up!"

"We—are—hauling—you—up," he repeated over the telephones, which were fortunately not yet reached by the sloshing water.

I felt my lifeline grow taut, and in a moment the deck dropped from beneath my feet. A compensating glow of satisfaction went through me as I realized that all was going well.

The water was growing lighter with each successive upward foot, and I felt the air expanding within my upper dress as the pressure fell off. Observing these indirect indications of upward progress—there now was no other measuring stick—it seemed as if I was standing still.

I reached out and felt for the lifeline. It was taut as a bow string and apparently was growing constantly tighter. I felt the lifeline twisted around my body, under my arms drawing and digging in deeply with increasing discomfort. Worse than that, I was losing my horizontal position. As the tension increased on the lifeline, it seemed to be pulling my hips and feet higher, and the water was running up into the back of my helmet.

I struggled violently in the three-dimensional nothing to right myself, but there was nothing on which I could grasp a purchase. The taut lifeline was jerking violently, conveying to me the pitch of the *Silver Heels*, and on each successive upward jerk, a surge of water rushed up into the helmet. Now the worst happened—my head and telephones momentarily went underwater.

I was soaked from head to foot. Worse than that, the phones were now just so much useless junk—it took only one drop of salt water to put them out of commission, and these had just been thoroughly dunked.

This would cut off our voice communication. A touch of the lifeline again reminded me that this was taut to almost breaking tension—how could I signal over this, now as still as a steel bar?

My immediate worry, however, was not to let that lifeline tension turn me upside down. If that should ever happen, the precious bubble of air in my helmet would instantly flow up into my feet (or out of the rent in the suit) and the water, inversely, would dump down into my helmet. I would then have no more chance of righting myself than would one of those potential saltcellars with the round lead base. This would be no minor matter to me—my head would be underwater—and apparently there was some difficulty in getting me up, for I was not moving. To avert the inversion, I reached again for the taut line and pulled my head toward it. As long as I could do this, I could keep myself upright, but it required an effort that now had me panting.

Perhaps I could get a signal through on my air hose. I reached for this, customarily made fast to the left stud of the breastplate front.

The hose was not there. I distinctly recalled their fastening it to this stud—I remembered feeling for it and finding it there on a number of occasions when I had checked my lines for freedom from fouling as I explored the wreckage of the *Dwight*. It must have torn loose from its lanyard, although it was apparently still connected to the helmet, for the blast of pulsating air was much evident.

With one hand, I groped frantically around for it, afraid to let go of the lifeline for fear of losing my equilibrium. Alternating, I tried the other hand. I was sloping off at about forty-five degrees in this world of water space, looking upward at the interior of a tiny globe of nothing emphatically broken only by the light-yellow lifeline. I could touch nothing. I tried my feet, kicking them back and around as far as they would bend but with no success whatsoever.

The strain on that lifeline was terrific and apparently was getting worse.

What if it should part?

It was not too heavy a rope—half inch (diameter) manila, which under ordinary conditions would lift a dozen divers with ample safety. Were the many gallons of water in the dress so heavy that the boys were afraid to strain it any further? We had many larger lines on deck but always preferred to use a smaller size for its corresponding lesser drag in the heavy tide currents.

Where was my hose? What were they going to do? Would that line snap? There certainly was nothing that I could do—it took every ounce of energy that I could summon just to keep my helmet above my feet, and that was vital! I was so cold that I frankly didn't care much what happened—my body was so numb that I now knew no feeling. I just hung on and waited, my hands now tiring, with much the same sense of desperation that a man hanging by his hands from a skyscraper scaffolding would have until such time as his fingers, numbed with exhaustion, relaxed their grip.

It was consoling to think that if there was something that could be done, Truman would do it.

Suddenly, the near fully taut lifeline limply sagged. I was falling like a lead brick. Down I plummeted as the water blackened before my window again. And then something soft came up to meet me—I was now on the soft oozy mud again off the wreck.

I now wheeled around and with a wild swing of my left hand was able to catch the air hose. I transferred it to my right hand, lifted my left, and passed it under that arm to its proper position. With the hose in my left hand and the lifeline in my right, I now sought to find out what the trouble had been. It was readily apparent.

The lifeline disappeared almost vertically upward, apparently directly to the *Silver Heels*. The air hose, supposedly coming from the same place, disappeared off to my left, making an angle of approximately thirty degrees with the bottom. It must have been fouled in the wreck.

I could neither go up nor down but could just wait. I wondered how much longer I could wait. I was so cold, I was afraid that I was going to lose consciousness—and that couldn't happen, since I was the only one who could free myself.

They were drawing harder and harder, the hose and lifeline again drawing taut as a banjo string. I hoped it wouldn't let go.

Suddenly everything went slack again, Truman again realizing that I must be caught and that I must have slack to clear myself. Again, I sank and soon my lead shoes dug into an oozy bottom.

Although our line signals were on the hose, I thought that I would try one on the lifeline in the hopes that they would realize the substitution. I gave one long pull, which means "OK" and was very pleased to receive the same signal back in confirmation. They understood.

The wreck, or whatever my hose was fouled up on, would be easy to find—all that I had to do was to follow the hose.

Pulling myself along on the tight black tube, I noticed the angle lessening and as soon as it approached the vertical, I found myself face to face with the slimy sheer wall of the hull. Closing off my exhaust valve, I again lightened myself but found that I could not float with the weight of water in my dress. However, I had the hose in my hand on which to climb, and up I went over the expansive freeboard of the vessel, arriving at her rail panting from the exertion.

The hose bent sharply over the rail and pointed off almost parallel with the deck into an apparent nothing but toward the troublesome snag. Stepping over, I pulled myself along over the deck until it led me to the source of trouble.

There, intertangled in a snarl of mast and rigging, was my shiny black hose. The other end was leading out of the snarl almost vertically upward toward the surface and was perfectly taut. Here was the mast for which I had been looking, lying over the debris on deck. From now on, I would know where it was.

It was only a few seconds' work to clear the hose, and then again, checking it and my lifeline for an apparent clear run to the surface, I gave the signal that I had hoped I would never have to use—four pulls—"Emergency, pull me right up to the surface without any decompression!"

I had overstayed my planned time of one hour and had a long decompression period—perhaps an hour and a half—but I couldn't stand that ice water another minute. The bends—anything—would be preferable, if I could be warm when I had them.

Truman's four-pull reply came back, and I knew that all was well. They all must have turned to on the line, for I started upward at a terrific pace, racing through the water like a speedboat. Less than a minute later, my hands were clutching the rungs of the diving ladder and my helmet was breaking the surface. Getting up that ladder alone on the side of the pitching and rolling boat was no less than a Herculean task. My 185 pounds plus the regular 200-pound diving suit, in themselves, were ordinarily no simple matter. Now, I had probably another 200 pounds of water to top it off, which, plus the heaves of the boat, made it next to an impossibility. I felt the welcome hands of Doc and Truman momentarily assuming some of the weight and, in a few minutes, I edged over the gunwale and lay belly-down on the deck.

This was not a particularly desirable position for sustained resting since the water now had run up into my helmet and my head was underwater. I pushed up with my hands, pointed to the faceplate, and instantly Truman opened it up to let out what must have been twenty-five gallons of ice water.

I knew that I would have to go back into the water again for decompression—but only thirty feet down for my first stage, where the water was considerably warmer than on the bottom. I asked Doc if he had some hot water on the stove. He didn't. He did have a huge pot of coffee brewing on the stove at boiling temperature. I asked him to get it and also an empty pitcher.

I drank as much of the hot coffee as I could. I then asked him to mix up all of the rest of the boiling brew with just enough cool water to take it to below scalding temperature. Doc eyed me curiously, wondering how I could drink all those gallons of coffee. I then asked him to pour the piping hot brew through the window of the diving helmet.

If ever there is any sensation that can transcend this for sheer ecstasy, it could only be in heaven. The hot liquid oozed around into my underwear and gave me my first reminder for some time that my body was still alive. Invigorated by the stimulation of the externally applied coffee, I was ready to go back and pay for my sins. Truman closed the window, and I slid overboard into the snarling seas.

Rather than rapidly sink to my thirty-foot stage, I hung on to the first submerged rung of the ladder and let the water run in through the rent in the suit. It would come in within a few minutes anyway, and I might as well select the comparatively warm surface waters for admission to the sacred portals within my water-soaked underwear. With the suit filled up to my neck again, I dropped down to thirty feet and then went about the long process of just waiting for eighty minutes to go by—twenty minutes at thirty feet, twenty-five minutes at twenty feet, and lastly thirty-five minutes at ten feet.

And then the most wonderful of all luxuries that man has ever invented—a dry towel. I wondered why I had ever chosen such a crazy profession as I shook with chills for the rest of the day. The boys connected our four anchor lines together and made them fast to a substantial buoy, and we shoved off for Cuttyhunk. The seas were still running heavily. Everything was in a mess after two rigorous days and a night. The gang were all half sick from exhaustion.

The quiet calm of Cuttyhunk Harbor was almost the revelation of Paradise. We all turned to and soon had the *Silver Heels* in shipshape order from the many neglected duties under the stress of the storm.

The next morning, we were all up at sunrise and rarin' to go—there was no gloom that couldn't be cured by a good night's sleep.

We all knew that conditions were not good. We had every large anchor we owned—four of them—on the four legs of the X and not enough rope left on board to anchor our vessel in anything but quiet water. This was poor policy, and we knew it. However, there was no anchor we could buy at Cuttyhunk. Likewise, we were short of anchor line, the depth of the *Dwight* and the currents running over her being far worse than our most dismal expectations. We had quite a list of things that we now needed and realized that sooner or later we would have to put in at some port where we could effect repairs and augment our equipment to fit the job.

Today, however, was not the day to go shopping. We had a minimum of everything we needed for today's dive—under no worse or no better conditions than yesterday's. Today, I would get into the cargo hatch.

With spirits high, we chugged out toward our buoy as I started to slip into the diving dress. Yesterday's mist had lifted, and we should have no trouble in picking her up. It seemed to me that we were getting into just about the right location. Truman, at the wheel, was apparently circling. He asked Doc to go aloft since possibly the heavy seas and ground swells were masking it from our view at water level.

Doc laboriously pulled himself up in the bos'n's chair along the swing-ing mast and finally perched in the cross trees, bracing himself by clinging to the topmost base. He carefully scanned the water through the full thirty points of the compass, checked it again, and then sang out, "No buoy in sight."

Our hearts sank!

Yesterday, in all of the confusion of my little accident, who had taken shore bearings?

20

CARGO

We didn't know where exactly to start searching, but we did know of one square where not to search. And we did know that she was someplace between that dragged area and Cuttyhunk, which was a lot more to start with than we had before.

Overboard went the drags and on labored the *Silver Heels* to their terrific back pull. We were retrogressing a few days, and we all were in the dumps about it.

Late in the afternoon, we agreed that it would be advisable to head back for our badly needed repairs, new anchors, and line. The only anchor we had was merely an overgrown skiff hook, which would be totally inadequate for any open-sea emergency. We hardly had enough line on board to hang Dave, recalling that there must be over two thousand feet of fine manila on our anchors, all of which were resting in nineteen fathoms of water someplace on the bottom of the Atlantic Ocean.

Now was the time—our drags needed repairs badly, and we had what might be no minor amount of this work ahead. It was late in the afternoon and the seas were running heavily still, so we hoisted all canvas and sailed for Vineyard Haven.

We secured around Martha's Vineyard and picked up four more 150-pound anchors, slightly rusty, used, but considerably reduced in price. We loaded up on rope, completed our repairs and changes, and headed back to sea.

Vose Greenough, who was summer vacationing in these parts, wanted to join us for a while, and of course, he was always welcome. He had done considerable sailing in these tricky waters and, unlike our last "expert," was not blind.

We headed for Cuttyhunk, breaking out every stitch of canvas plus our engine to breast for spring tides running through Vineyard Sound. This island, for so long our home, almost looked good to us this time with our spirits again running high.

We were finishing up our usual evening feast, a three-course dinner consisting of lobster cocktail for a start, Lobster a la Newberg as the entrée, and then lobster salad—plus anything to drink that we wanted as long as it was Doc's quick mix of rejuvenated powdered milk. A fisherman ahoyed us, came on board, and said that he had found our buoy and did we want it. I had my name, "Nohl," painted in big letters on the fifty-five-gallon drum to establish claim on the wreck.

We scurried over to his boat and were pleased to recognize and welcome back our much-needed marker.

Truman cried, "Look, this line has been cut with a knife—that didn't chafe through." The fisherman said that he had not cut it—that was just the way he had fished it from the water.

We don't know to this day who cut it loose. We did learn a little later that both Captain King and Captain Carmichael were still alive and in this country—and that they were watching what we were doing.

The next day, we shoved off at sunrise and started dragging, this time not with the spreader drags but with three grapnels trailed along the bottom behind the *Silver Heels*. We thought that it would be easier to snag them in our huge X of manila rope than to pick up the wreck itself—and we were first interested in retrieving any part of this spiderlike network so that we could pull it up and anchor to it.

Late that day, the buoy caught on something and slowly eased us down to a stop. Uncle Hal, throwing his clutch into reverse, gave her sternboard as we kept the grapnel line taut, and we backed to get over the snag. As we did so, we pulled up—and soon we had hold of our lost anchor line.

We did not know which was to pass to get to the apex of our X but on fifty-fifty chance guessed right and soon had the ring to which all found lines were attached. We grasped the fifth line to this same ring and pulled it in—the buoy line.

Its end was not frazzled as would be a chafed line. It was cut clean as a whistle. There had been some monkey business going on.

We refastened our buoy to it and, with another line fitted with a swivel so that we could swing to the tides, made fast the *Silver Heels*.

Vose was with us that day, and I knew that he was anxious to take a dip. Although I was veritably itching to get down into the cargo hold to see what Fortune had spun for us, I might never have been in this business in the first place if it hadn't been for him; and thus, he was cordially invited to take the first dive this morning.

He had a bad day. We were in the full fury of spring tides, and the current was rushing by like an overzealous mountain stream. These currents, scrambling pell-mell over the oozy bottom, had churned the water into what momentarily might have seemed a muddy river. The *Silver Heels* was heaving heavily to her hawsers as the giant seas swept impetuously beneath her hull.

Almost half a dozen years had slipped by since Vose had helped me take my first dive in a regulation suit through the ice covering of Walden Pond. I had, in the meantime, bought all of his equipment and was now eager to watch my former teacher through my now long-experienced eyes. He had not forgotten a bit of it and, in his usual thorough manner, checked through every detail and entered the water like a first-line professional.

Vose did get himself into a bit of a jam on that dive, although I am confident that it was through no fault of his own. The descending line, still fast to the port rail adjacent to the cargo hatch, was almost vertical up from the wreck to the diving ladder. With the heavy ground swell, this would alternately tighten up and slacken off as our vessel rose and fell to the passing mountains of water. Afraid that, if it should ever draw completely taut, something would have to give—the *Dwight* rail, the surface cleat, or the line itself—we dropped off ample slack and then just a little more to make sure.

This, I am sure, was good seamanship, but to Vose sliding down that rope for a descending trolley, it was a nightmare, as the line jerked up and down like a yo-yo string. As he approached the wreck in that day's muddy waters, the coiling and uncoiling snarls of slack were hungry traps. Exactly what happened neither we nor mud-blinded Vose could see, but soon after touching his shoes on solid deck and noting that visibility was about six inches, he gave the signal—three pulls—to be hauled up.

The phones, as was more often than not the case, were out of commission, still recovering from their recent saltwater bath, and so we were back to our well familiar line signals. We pulled him up but apparently lifted him just about a fathom when the line drew taut as a bow string, indicating that he was fouled. A couple of jerks failed to break him loose, and we slacked off and let him drop back down to the deck. A few minutes later, another three-jerk signal came through, and he came up a second time for a solitary fathom, only to draw up taut again.

Down he went, up he came, down he went, and up he came, not over a fathom each time. The poor fellow was stuck down there but obviously still functioning, since his signals were coming through and apparently to an intelligent plan.

After about six repetitions of this cycle, we were beginning to seriously wonder what we should do. The diver, however, is king, and as long as his signals were coming through, we knew we had better just pull him up and down as he requested his allotted fathom.

The next time, to our pleasant surprise, up he came perfectly free and easy. As his dress later emerged from the water, we began to see what had happened. Vose was wrapped in a tight coil of manila rope as if he had been trussed and bound by escaping burglars.

After his helmet came off, he explained that he had been entangled in such a horrible snarl and in such dark water that he realized that not in a thousand years of frantic tugging could he disentangle himself. He, as we always required, carried a sheath knife on his belt with which he could cut himself loose from foreign entanglements. However, in the dark water a slight problem had arisen as to which was his own lifeline and which was the fouled descending line. They were indistinguishable in appearance, both unfortunately having come from the same original coil of rope. Apparently, he had guessed the right one, since there was Vose on the end of the rope, alive and kicking.

As glad as we were to note that, we were all just a little discouraged when we realized that in cutting the descending line, he had now lost the wreck for us. We had to start all over again.

It would not be all over again, however, since we were still anchored in her approximate position, but we would have to put out in the dory with grapnel and line and pick her up again so that on the next dip I could land on at least some part of her deck.

Doc and Truman shoved off and in about fifteen minutes sang out, "Wreck below!" as they prepared to lead the upper end of the line over to the *Silver Heels*.

In the meantime, I dressed and was ready to go overboard by the time they were back and all lines made fast.

Purportedly to save time, but actually "just for the hell of it," each day I had been trying to break my record in hitting the bottom. On that day, Truman told me that he had timed me at forty-eight seconds, which I had made by a combination of deflating my dress to almost a "squeeze" state of collapse (to minimize buoyancy) and by pulling myself downward on the descending line hand over hand to aid the none too ambitious gravity acting in this watery world.

The tide was now slack according to the tables, but that just meant a minimum current—it was always running in this location. However, the water had cleared considerably from the heavy cloudiness in which Vose had been, and I found that my visibility was at least three feet.

Crouching down to include the deck in the visible sphere, I crawled around on the slimy structure in several trial directions until I ran into a rail. Feeling my way along this for a few feet each way, I decided that it must be her taffrail. This seemed evident from the rounded curve that it assumed as it followed the lines of what could only be her stern.

Crawling forward, carrying the grapnel which had caught on a massive timber of her wreckage, I moved in the direction that I thought—but mostly hoped—would be forward, in the hope that I could work my way toward the cargo hatch.

The rest of that hour accomplished nothing. I soon found myself in an almost impenetrable mass of wreckage—huge timbers, splintered decking, wrenched iron piping, frazzled wire rope, and broken hunks of engine castings. The entire wreck was covered with a slippery blanket of brownish green slime cloaking everything alike from ready identification. Rubbing this off here and there, I found a strange assortment of rusty tools, huge bolts, and loose deck fittings—almost anything that a ship would carry.

Apparently, I was now in her engine room. Even more apparently, here was where the explosion, which Roland Snow had heard, had taken place on that early April morning in 1923. This was not too healthy a place for a diver with ambitions of growing to be an old man, but I was anxious to get a glimpse of this portion of the wreck sooner or later. I also wanted to pass through it so that I could reach the cargo hold to refasten the descending line to the rail from where Vose had cut it.

Under the weight of my lead boots, the wreckage crumbled beneath me, and my hands and feet slid off the slimy slippery timbers. I lightened my suit as much as I dared and proceeded with the caution of a mountain climber scaling a sheer rock precipice, firmly feeling each new foothold for assured security. With every move, I stopped and rechecked my lines to make certain that they were leading clearly upward into the nothing above me that represented the world from which I had come. In a moment, I was glad that I had done this.

I pricked up my ears. Something was wrong!

With the same filtering instinct that a mother must use to discern her babies' disturbed breathing in the night, I was suddenly subconsciously aware that there was something wrong with the incoming air. The roaring din of the entering air and the steady rhythm of the compressor were ordinarily almost unheard by my paralyzed ears—until there was as much as an indistinguishable but real change in their tempo. The monotonous drone, like a long-endured buzz, was only conspicuous by its absence or change.

It was now distinctly falling off. Poonk-poonk-poonk–poonk–poonk–poonk–poonk—poonk—poonk—poonk—poonk——poonk——poonk——poonk——poonk———poonk———poonk———poonk——————and then silence.

I quickly grabbed for my exhaust valve and tightened it down to preserve as much of the suit's buoyancy as possible. Fumbling again in the dim light, I reached my air hose and lifeline and found that they were all clear. I checked my feet, legs, and crotch to make sure that I was not snagged in the snarl beneath me. A signal would be coming through momentarily, since the boys on deck would be as conscious as I of the compressor failure. I then took in a few feet of slack on the hose to hold and await the signal as I again rechecked everything.

In a few seconds down it came—one long pull of the hose, another pull, a third, and then a fourth.

This was not, "We are going to pull you up." This was, "Emergency!"

It was a warm gratifying feeling to feel those pulls coming through so firmly and assuredly, reminding evidence in this dim watery world that my keepers were caring for me. I gave the signal back in confirmation, thoroughly sure that I was all clear.

As I barely completed the end of the fourth pull of my answer, I felt the lift line drawing around my waist and the few pounds of weight that I still retained on the wreck disappearing. In a moment, I was on my way upward at a rapid rate.

The water was brightening so fast that it hurt my eyes. I felt the strange sensation of my chest expanding without inhalation as the air in it and in the dress expanded with the decreasing pressure.

Twisting in the water, as I realized that I was approaching the surface, I saw the bottom rungs of the ladder a few feet before me and reached over to grasp them and help myself up. I must have been on the bottom for most of my allotted hour and theoretically had an hour's decompression ahead of me. This is a situation that might at any time face any diver on a deep bottom—what to do? The air had stopped. Do you stay on the bottom to suffocate or go up and get the bends?

I didn't have the bends yet—this, strangely enough, was usually a malady that set in some time after the diver had come out of the water and was something that I would have to look forward to in anytime from five minutes to five hours.

Doc opened my window, and Truman came over and advised me that the air compressor had stopped. I was aware of that. He explained that the engine had just stopped for no apparent reason and that Uncle Hal was frantically trying to restart it, but with no success so far nor any sign of life from it.

This was really no serious emergency in this case, since we were prepared for it. Doc had out an open-end wrench and was already disconnecting my hose from the compressor outlet to transfer it to our hand pump. In a hurried few seconds, the job was done and he leaned on the pump to start the air on its way.

This pump was only good for thirty feet of water, but that was all that I needed, since the balance of the dive would be in midwater decompression and not on the *Dwight*. Truman dropped me down to that level where I hung for fifteen minutes; then pulled me up to twenty feet where I hung for another twenty-five minutes; and then up to the last stage at ten feet where I hung for thirty-five minutes to finish the dive. This type of decompression, going up to the surface from deep water and then back down, is not particularly recommended, but it has never failed me in an emergency. I had not even a minor trace of that dreaded disease.

We were still right back where we started—the loose grapnel came up, and we had no line to the wreck. Tomorrow, certainly, I thought I could find that cargo hatch and at least get back to where we were a week ago— our farthest point of advance.

It was still early enough to fish for the wreck and have everything ready for tomorrow, which Doc and Truman did. The tide was now flooding again, and they struggled madly in the milling stream. However, they soon had her snagged and decided to bend the grapnel line to the buoy. We decided that this would catch the wreck in several places, and in the morning if I landed on her in the wrong place, I would simply come up and try another trolley. Tonight, we would plant three or four on her.

They tossed the buoyant buoy over, but it sank like a sounding lead. Although it was a five-gallon can sealed hermetically shut with a solder, that tide was running so swiftly that it towed the buoy under as would an oversized fish pull down an undersized cork on a fishing line. Our buoy was gone, but as soon as the tide slacked, it would again reappear—we hoped—if it hadn't been drawn so deep that the pressure had in the meantime crushed it. Seeing this, we decided to wait to put any more buoys out and prepared to go back to Cuttyhunk.

We had forgotten—our engine was on the fritz. Uncle Hal had been working furiously on it but as yet had not brought out as much as a cough. We were rolling and pitching so violently to our hard-biting anchors that none of us particularly fancied the thought of spending the night out here in the boat. Bucking like a wild horse, it was particularly difficult for Uncle Hal to work on the engine, and so we decided to sail in.

We would under no condition attempt to sail through that narrow passage between the islands into the harbor, but if we could get near to its mouth by canvas, we could anchor and would undoubtedly be able to pick up a tow through the pass. We hoisted jib and main and prepared to cast off the mooring line.

This would be a tricky operation, and we knew it. Beneath us was a maze of four anchor lines, buoy line, and grapnel line, all in a position that we could not exactly determine in the swirling waters. We would simply have to cast off and drift through them. Truman threw over our mooring line, and we seemed to leap downwind like a spirited horse finally free from the bit. Our bow swung off, the jib and main filled, and we felt the vessel take on forward way. She started off smartly but then somehow seemed to draw up short, all canvas drawing but our bow falling farther and farther off. We were swinging violently around with sheets well drawn in but apparently with no forward movement and no steerage. As we swung our beam into the wind, it was necessary to quickly slack out on the sheet to avoid a dangerous heel. Apparently, we were spinning around on some point under the stern, probably our rudder or propeller being caught on the anchor lines. As we swung with our quarter to windward, we were in danger of a bad jibe, and we called to Doc to quickly drop the main peak. Down came the throat subsequently, and in turn we lowered all of our canvas. We were now riding to our stern, caught in some mysterious way under the counter.

With the pike pole, we fished around beneath the boat and found that we were trapped on our own maze of converging anchor lines. We jerked, pushed, and pulled, but nothing we did could dislodge the tangle.

This was something that we had long realized might happen sooner or later, but we knew in an emergency, we could simply go under our hull and unsnag it. After all, we were a diving boat. However, today our engine was broken down, and our only source of air was the hand pump. I had a ripping headache from my last hour with it, which had not offered enough to keep the helmet from being foul.

I was cold, too, as I usually was after coming up from the icy depths, probably more than anything else from a two-hour exposure of bare hands and wrists. I did not particularly fancy crawling into the clammy gear again. Perhaps I was rationalizing, but I expressed hopes that for that little job it would be more practical just to swim under the overhang and clear the fouled lines.

Uncle Hal had long claimed a prowess at underwater swimming, and so now we called his suspected bluff. Overboard he went, disappearing under the stern, apparently capable of staying submerged, for we thought he was never coming up. When he did, it was with a report that he couldn't seem to see enough in the dark water under the shadow of the hull. He tried it again with appropriate resting spaces between but to no avail.

Doc next tried but after a few attempts arrived with chattering teeth at the same conclusion.

It was now getting so late that if we were to make the pass before dark, whatever was to be done would have to be done immediately, and so I decided to try it. However, my attempt was more futile than each of theirs, since I was so cold that I could not stand the frigid water, and my hangover from the last dive seemed to result in a serious inroad on my air capacity. I could see, with my own little trick of establishing vision underwater by squinting my eyes, that the prop was badly fouled with turn after turn about its hub and blades. We were riding to these very lines, and they were so taut that working with them or untwisting them was virtually impossible. The only thing that I could do was to make certain that the rope was not now passing over the propeller blade, which might cut it or itself be bent.

It became apparent that we would again sleep over our treasure ship, only one hundred and twenty feet from a fortune that we hoped would soon be ours.

Because of the danger of our prop cutting through the line and setting us adrift in the inky black night, we stood continuous watches until sunrise—we had no engine, and that was not a pretty prospect.

It was a jerky night, the *Silver Heels* laboring under a new motion, anchored by her propeller. In the morning, we all were in good spirits again. After a dandy breakfast—we were in no hurry, since it didn't look as if we could dive that day anyway with our engine out of commission—we went about our tasks. Uncle Hal returned to his engine, Doc washed the dishes, including an accumulation of the last few days, and Truman helped me into the diving gear.

I was all ready to drop overboard, Truman having passed a loop under the stern that I could hang to, when unexpectedly the *Silver Heels* quivered and shook. The engine was running.

Uncle Hal appeared with a big grin from the engine room and announced that our only difficulty had been water in the carburetor. He had cleaned it out, and she was purring like a sewing machine. I was happy to find that I could now go down on the compressor rather than on the hand pump, and soon Truman made the changeover.

I grabbed my trolley and slid down under the counter and soon arrived face to face with the propeller. Slipping my left arm over the rudder support, I hung there on my armpit and found that in this way I had both arms free to work.

I soon realized, however, that I would need more arms than an octopus to do this job. The rope, slanting downward into the deep green nothing below, was so taut that it might as well have been a steel bar. Pulling as hard as I could, I was not able, at least in this position, to get the slack that would be necessary to take the turns off the wheel. I could cut it and free the boat, but there would go our badly needed anchor or anchors and equally precious attached lines. There must be another way to do this.

I crawled back on the trolley loop to the diving ladder and emerged high enough to get my window over the gunwale. Truman opened it, and I explained my little plan.

He listened, heartily approved, and down I went, carrying with me the end of a second line. Crawling back along my trolley loop, I arrived at the prop, transferred to the anchor line to which we were riding, and then slid down about fifty feet, the line still being as taut as a bowstring. On this I took a half hitch, another half hitch, and then a clove hitch. Crawling up the line again, I returned to the propeller. Truman, noting this, as we had agreed, borrowed the boys for a minute and hauled in on the new line. This now assumed the strain, and the line to the propeller now went all slack in my hands. It was now just a matter of a few seconds to unfoul the turns, transfer to the trolley loop, and shuttle back to the ladder.

With engine now running, Truman kicked her ahead, and in a moment, we were back at our buoy and moored fast to our bow cleat as we should be in proper seamanlike style.

All night long, we had been trying to get back to Cuttyhunk. We could now go.

"But, why should we go?" I thought. Here I was, in the water up to my hips, in the diving suit; the engine was running beautifully; and the sea was even moderate. It was a good diving day—as good as we had seen all summer.

The grapnel was still caught on some unknown portion of the wreck. Its buoy, which had last night towed under in the flood tide, was now again afloat, dented in slightly by the pressure but not leaking a drop, and Doc had rowed over to pick it up.

With the line in my hand, I slid down and hit the wreck in fifty seconds. The water was almost slack and as clear as it had ever been. The hook had apparently caught on a section of rail, but not knowing on which side of the wreck I was, I selected a direction and followed it along. In a moment, I arrived at the fo'c's'le head and realized that I had landed on her starboard side and now knew exactly where I was. Retracing my steps aft, I took a bowline on the bight around the rail where the grapnel had fouled. This morning, our luck was running almost perfectly. We were now up to exactly the same point of progress at which we were a week ago.

Starting amidships, I soon found my calculations correct as I approached the hatch coaming. It was early in the dive—I had my full time—and I looked over the slime-covered edge into the black hole below and realized with a swelling of excitement that within this hour, God and the air compressor willing, I would touch our pot of gold at the end of the rainbow.

Edging along the coaming over the deep brown mossy deck, I found a ladder leading into the abyss below and swung my lead boot over the side to start down. The compressor was purring like a kitten; my lines were clear; I could see almost six feet, which was heaven down here; and I felt like a million dollars, physically and spiritually.

The water darkened rapidly as I climbed on down the slippery rungs, my six-foot radius of visibility decreasing decidedly. And then, I struck something solid and soon was standing on a firm base.

Bending down, I found that it was not a 'tween deck. It was not her bottom. It was not wreckage.

I was standing on a huge pile of bottles.

21

THE TREASURE

I lay flat on my belly like a parched desert traveler arriving at a long-envisioned waterhole. In this case, it was more to try to see the nature of what I had discovered than to imbibe of its contents, which would, under the conditions, have been a bit impractical.

They were whiskey bottles—the familiar Black & White shape that I had, for long months, eyed with no casual interest in liquor store windows.

The light had been miserable before but was now worse. A fine oozy silt had settled over everything, and my digging had immediately stirred up a cloud that would not clear nor settle in the quiet sheltered waters of the deep hold. It wouldn't be fair to say that I was in total darkness. I could press my finger against the outside of the helmet glass and barely see a flattened fingerprint. However, an inch away, my finger was totally indiscernible; I imagined I would quickly lose it if it were not connected directly to me.

My attempts at examining the bottles were as futile as trying to identify the proverbial black cat in the coal bin at midnight. However, I could send some up.

Truman and I had arranged a special signal for this long-awaited moment—two shorts, two longs, and two shorts on the air hose, all of which would simply mean, "Send down the net." Weighted with a hunk of iron, clipped to the lifeline with a snap clamp, down it came directly to my waiting hand. I unsnapped the net and started in the inky murk to load the precious bottles. I couldn't see a thing but certainly didn't have to look far for them, for they were everyplace beneath me. I could still think and dream; as I placed each one into the container, I thought of the sizable price tag I had so often seen next to the familiar Black & White fifths in the stores.

Afraid to load it too full for fear of spilling, I signaled when I had what would probably be about two hundred dollars' worth of the "liquid gold" in the basket. In a moment, I felt the line draw taut as the boys started to draw it to the surface. Guiding it as high as I could reach, I felt it finally disappear into the black nothing above me. I now simply had to wait for the next empty basket to return.

Later, I thought, we would use bigger nets. We could also have two in operation so that I could be loading one as they hoisted and unloaded the other. Perhaps a little light would help down here, also. We might even drop a centrifugal pump intake down here so we could pump out the murky water and improve visibility. With a few refinements, this would be a snap.

In a few moments, back came the empty container, and I again started loading it. While waiting, I had built up a little pile of bottles in front of me, and it was only a matter of a minute or two before I was ready to signal the "Hoist away" for net number two.

Each one of these loads would be worth at least a couple of hundred dollars.

Truman was trying over and over again to call me on the phones. However, these instruments, after their last saltwater bath, had never been the same. On some days, they would barely get the signals through—on others, there was just a faint mumble. I tried, by turning off the air, to get his signal but finally gave up. He was trying to say something, but I didn't have time to put in the necessary work to hear it. If there were more important instructions, we had rope signals to cover every exigency.

Quickly, in the inky gloom, I continued to load bottles. This was the day of which I had dreamed for years.

Finally, the three-pull signal came through. I filled the last basket, and then, just to get a few more, filled the two big pockets in my chafing pants. I still had to hang a full hour at various levels in midwater to decompress. That was the longest hour of my life. I was cold, and up on deck was a drink—a very special drink of scotch whiskey—waiting for me. Why shouldn't I have a little "warmer-upper"—we had a third of a million dollars' worth of it!

Nohl handing bottles
to the crew.

For the past couple of hours, the boys up on the deck had been experiencing a variety of emotions. To them, for these long months each wave had looked alike. Regardless of where we had been or what we had been doing—dragging an unseen bottom or anchoring over the rotting hulk of our mystery ship and its fabulous treasure—each patch of water appeared no different, except for the reports I had brought back, from the next one. I had gone down into that monotonous surface over and over again, but to their eyes as I emerged, there was little change in my appearance other than the added dripping wetness. But now they had dropped a net down into that blue-green nothing and up it had come loaded with scores of bottles of whiskey. All of the months of dismal discouragement now seemed justified. We had been right. By sticking through it all, we had finally received our reward. Here it was, coming out of the water.

Celebration was in order! And what could be more appropriate with which to celebrate than our own scotch—ice cold out of the dark recesses of Davy Jones's Locker.

They had brought out the glasses and all assembled for this historic, momentous drink. Truman had cleaned off the marine growth on the outside of the first bottle, to prepare to pull off the cork. Yet, there was no cork! The bottle was partially sealed with a cork-shaped puttylike mixture of slime and mud. Turning it upside down, out came a mixture of salt water and slime.

The other bottles were about the same. The contents of a couple of them were not completely displaced by the seawater—there was enough scotch left to impart a definite flavor. What a drink that was! Wow! Salt water flavored with scotch!

By the time I emerged from the water, climbed the ladder, and waited for my helmet, ready to enjoy the forthcoming big moment of my young life, Doc, as usual, had a pot of hot tea waiting for me.

22

Fo'c's'le Head

The next few weeks were routine diving. Everything was in our favor—the seas were moderate, our anchors were holding, and the descending line was always there, now secure to the *John Dwight*'s main beam, to lead me directly into the cargo hatch.

In those weeks, I dug down into the cargo of bottles like a groundhog, finally finding myself in a hole deeper than I was tall. From the bottom of this hole, I fished out the deepest bottle that I could grasp, but this was in the same condition as all the others.

Apparently, as we later learned, this bootleg shipment had had a very slight variation from any other scotch bottles of those days—a matter of such small detail, and yet one that explained the discrepancy between our cargo and other cases of scotch that had been salvaged from even greater depths since Prohibition.

These bottles, known to be for contraband sale, had not been filled quite as full as would have been the legitimate containers. Perhaps it was the difference of two or three teaspoonfuls in a fifth of a gallon bottle. This slight drop in level had resulted in an air space, which under normal conditions would have meant the loss of just the corresponding two or three teaspoonfuls of whiskey—but to this load, it had meant a place for the corks to go.

Under the pressure of the waters, the corks had simply been pushed in. In a properly filled bottle, the long cork could only move a harmless fraction of an inch before it compressed the trapped air to the pressure of the water, after which all would be in harmony for almost an indefinite period of time. In our case, the corks had been forced down into the bottle before that pressure equilibrium had been established.

Two teaspoonfuls of scotch had cost us $350,000.

We didn't give up easily, however. After digging my hole, I decided it might be wise to try other locations, in the hope that some of the bottles might have been fuller or for some reason had not lost their corks. I ripped off some of her deck planking in several places to get down into the hold in various spots but found the same thing. Dropping over the side of the ship, I tore through her side at about the turn of the bilge, which had been swept clean of settling silt by the fast tide currents, and extracted sample bottles here, but I never found one with more than a taste of scotch in the salt seawater. What a vile drink that was to sample!

At various points, I ran into her cargo, which we had known was Frontenac ale, and on which we had never pinned much hope because of its moderate bulk and comparatively low value. These bottles came out clean as a whistle—the Crown caps on them had been eaten off as nicely as if taken off by a bottle opener. This, apparently, was a galvanic action in the salt water—the metal of the cap interacting in a battery-like action with some other less active metal nearby. Likewise, the metal hoops on the barrels had been completely eaten away, leaving only a collection of loose barrel staves. These barrels had not carried bulk liquor but had been planted there merely to represent the alleged legitimate flour cargo, actually having been filled with loose scotch bottles packed over with loose flour. There was still evidence of the flour, but not exactly in good usable condition.

There was still a little matter of $225,000 in cash that definitely was in the wreck. My hope of finding this could only be described as negligible.

The explosion had torn up the after portion of the ship into just a pile of loose wreckage. Her entire superstructure there had been blown up, her main deck ripped out, and the hull so gutted that this mass of wreckage would be a summer's work to entangle. This, if ever such existed, was a treacherous place for a diver to work with the odds of deep water, low visibility, and strong currents against him.

The money, in one-thousand-dollar bills, was probably in no more pretentious a container than an empty coffee can. Bootleggers just didn't use safes. The money had probably been hidden in some remote corner, and finding it would require tearing the vessel apart stick by stick.

Covered with marine slime, everything looked alike, and I might easily have stepped on the money or touched it in my various accidental visits to the mutilated stern.

There was one rational possibility—it might be hidden in the fo'c's'le. This portion of the ship was in good condition, and I was eager to visit it—if it did not contain cash, I might at least run into something very interesting. After all, I was in this business for the adventure of it. In the forward quarters, almost anything might happen.

It did.

A few days later, I slid down the descending line but this time, rather than following it into the cargo hold, got off on deck and then walked forward. Finding the rise of the fo'c's'le head, I moved to starboard and shortly found the companionway leading to the quarters below.

I probably earlier would have ventured down into those quarters (where perhaps still were reposing the bodies of the other eight men in the crew) if I had not rationalized that the cargo examination was our first consideration. There was actually another good reason for delaying this: the passageway down into this portion of the ship was completely blocked by a pile of wreckage—a grim tangle of masts, spars, timbers, anchor chain, anchors, and miscellaneous rigging.

Of more concern than anything, her main anchor was lying under the rest of this wreckage directly over the companionway.

Companionways are characteristically designed with very little consideration for spaciousness—always awkward for the excessively fat or tall. With the bulk of the diving suit over my not exactly diminutive frame, this passage, after being cleared of the wreckage, would not be an impossibility to navigate, but it would involve a bit of planning and wiggling.

It was now time to take the proverbial bull by the horns and get started with the job. Laboriously clearing away the debris hunk by hunk, I found that by the end of the dive it was completely ready except for the removal of the anchor. Next time, I would tackle that.

The enormous weight of the huge hook could be lifted only by the full flotation force of the *Silver Heels*. Fastening a heavy line around her hull and connecting the lower end to the *Dwight*'s anchor, the boys could quickly take up the slack as she dove into an oncoming wave. Then, with all lines taut at the lowest part of the cycle, as the rising sea tended to lift her, something would have to happen—either the anchor would lift, the rope would break, or the *Silver Heels* would sink.

We tried it and were pleased to see the first of these possibilities take place. My respect for the power of those seas mounted even further as I saw those tons of iron snatched up and dropped like a little toy. The anchor landed with its shank just clear of the companionway entrance, not completely out of the way but enough to allow me to squeeze by.

The next job was to take inventory of all of the protrusions on the diving suit and make a recheck on both air hose and lifeline. The telephones, as usual, weren't working very well, and I wanted to make sure that the pull signals would be clear at least to this point. Looking up, I could see the lines disappearing into the bright bluish void above. I was ready to go.

It took a good bit of squirming to ease the many helmet protuberances through the small passageway, but soon I found myself free and clear within the interior of the ship. Looking back, I again checked my lines and saw them still shooting upward after passing over the massive anchor shank poised at the edge of the entrance.

It took a minute to accustom my eyes to the dark interior. As my pupils dilated, there before me loomed up out of the nothing the outlines of the fo'c's'le. I could see the bunks in double tiers stretching forward and converging toward the bow. Almost feeling my way along, I picked up hunks of soggy cotton from the badly disintegrated mattresses, pieces of rotted blankets, and a slime-covered shaving mug. Upon rubbing my fingers, it gleamed against the dark background, the white glistening china apparently totally indifferent to the years of waterlogging which had affected everything else. A brass safety razor seemed to be in perfect condition, but the steel blade within crumpled in my hand as I curiously opened it up. A flat piece of moss-covered something or other turned out to be a mirror, which still reflected my gargantuan form as I scraped a path on it with my thumb. Another similarly shaped object proved to be a photograph in a frame—the wood spongy, the glass cracked, and the picture itself rotted to jelly-like consistency—whether of a battleship or a best girl, no one would ever know.

There was a kerosene lantern, so spongy that I could crumble the metal in my hand but with the glass chimney as good as new. There was a pair of overall pants, apparently still in usable condition, but of somewhat limited salvage value.

Starting with the port after bunk, I made a routine examination of all possible places where the money could have been hidden, feeling for any loose board in her side and top ceiling behind which a coffee can of cash might have been tucked.

Working forward, I glanced down and saw a weird sight. On the deck beneath my boots were the remains of one of the men.

He was wearing a regulation navy-type pea jacket, dungarees, and apparently beneath it all and holding things together, a suit of woolen underwear. In almost perfect juxtaposition, protruding from the sleeve ends and collar were the slime-covered bones and skull of the murdered man. A kick with my foot over his legs and body indicated that there was nothing inside but the skeletal remains—picked clean by the cannibal fish, who even now, crowded into the cabin, were watching closely every move that I made. Later, I would examine him more carefully, but now I was anxious to finish the survey as my time on the bottom was rapidly drawing to a close. He wasn't going anyplace.

Suddenly, I subconsciously sensed that the smooth rhythm of the air compressor was slowing down. It seemed to speed up momentarily, then again slow down to an alarmingly low beat, then gasp along for a few turns before giving out the final death rattle. There was now a frightening silence in this deep dark tomb, the final resting place of many adventurers before me.

There was no time to lose. I had the usual few minutes of consciousness in which to perform the delicate operation of crawling back through that companionway. Once on deck, I could pass out at my pleasure.

I could make it all right. If I hadn't thought so, I would never have entered this tempting trap.

Again, I broke out with my little poem, "I'm out of air and I don't care—I'm out of air and I don't care." I knew I must do my duties with no increase in emotional level whatsoever. The moment I got excited would be the moment I was licked.

My lines were somewhat snarled about the various items of fo'c's'le furniture, and so I methodically went about feeling them out and getting everything clear back to the companionway. There was quite an accumulation of snags down here. I was at the extreme forward end of the space, so I well knew that I would spend the bulk of the remaining time merely getting ready to squirm out of the companionway.

Here is where the telephones would have been of tremendous value. We had had so much trouble with them that they were now completely out of the helmet and stored away in a dry place awaiting the first opportunity for an overhaul and rebuilding.

A signal was coming through on the line, but it was so tightly snubbed as it entered the fo'c's'le that it was impossible for me to ascertain exactly what it was—like hearing in the night the distant striking of a clock but not being sure whether the first chime had been heard. But it felt like four pulls, the emergency signal, and it very well could be with no more air coming down.

I was still far enough in that it was impossible for me to get any signal through nor did I want to waste the precious time that this would take. I just disregarded their obvious signal until I could get to the opening and then make clear my wishes.

It seemed as if a second series of pulls were feebly coming to me, but I still could do little in confirmation. One difficulty with the diving business is that the men up on deck are always much more concerned about the dying diver than he is himself. I can understand that feeling. In the few times that I have tended for my friends, I have deeply sensed the feeling of responsibility that seems to weigh on the tender. In his hands, he is holding the only two threads of connection between the world of air and his charge below. Because the diver is totally unseen in a watery world that instinctively spells death, and because the imagination runs so riotously on that which cannot be perceived by the eye, the temptation is always to be overzealous.

Now, if ever, with the air compressor broken down and with no response to the emergency signal, they succumbed to the natural temptation to "do something," and the only thing that could be done was to pull.

It was understood that they must never pull until they received my answer in return—but this, they thought, was different. I had no air—I must be unconscious, and although the rule was clear, rules must always be tempered by judiciousness.

It seems hard for most laymen to understand that the shutting off of the air supply does not mean instant death to the diver. The air was still sweet, and if I kept cool, the first couple of minutes would mean no more distress than that commonly occasioned by holding my breath in a bucket of water for that time.

Anyway, after no answer to the third emergency signal, they started hauling. Encouraged by the slight slippage as the lines snubbed over the shank of the anchor, they kept on pulling harder and harder. They were now dragging me through the wreckage-filled quarters like laundry being pulled through a ringer. Something had to give to that relentless drag, and on I went toward the companionway entrance.

With the lifeline tied around my waist, I was not being towed head-on like a boat but was spread-eagled and being pulled along perpendicular to my path of progress. I was frantically fighting to clutch something solid, but everything within grasp seemed to crumble at my touch. I made a last frantic effort to push myself back so that I could steer my helmet through, but they were playing me like a sailfish, keeping those lines as taut as an elastic.

And then the worst happened. There was a terrific crash, and the huge anchor slid down the companion slides and settled right over the middle of the exit. The tension in my lines, rubbing over the anchor shank, had upset its precarious balance.

If I had had a full hour ahead of me to figure out a way to move back that anchor, I would have had plenty to be concerned about. But now there were only a few fleeting seconds left. I remember the prodigious efforts it had required to move it before with the entire freedom of the deck, but now from inside of my little dungeon, the prospect was not too pleasing.

I was really in a jam. I thought of my friends up there and what they must be thinking—my friends only 110 feet above me but as far separated by actuality as if by interstellar space. Between us was only water, a substance in which man was never meant to live—a substance which I had penetrated in an artificial envelope of rubber, canvas, copper, and glass. I had, for all of these years, defied the jealous gods of the sea, but perhaps now it was their turn to defy me. But if I ever did get out of there, I'd find out what was wrong, fix it, and come right back.

I was panting frantically. Everything was getting black. "I have no air and I don't care," I said again. And that was all I could remember.

I seemed to hear the distant peals of a bell. It was ringing rapidly. Where was I? Where was that bell? No, it wasn't a bell. It was a pfoo—pfoo—pfoo—pfoo—pfoo. I opened my eyes. I could see a rectangle of light above me. I was lying on my back on the floor of the fo'c's'le. That was the air compressor. They had it working again. My lines were slack. They must have realized that I had been jammed tight by their pulling.

I had a splitting headache, but that didn't matter. Now my strength was coming back, and I now looked forward to the problem ahead of me—moving the anchor.

Crawling up the cabin ladder, I peered out on deck to study the situation. I made an effort to budge it with my hands but soon found that would be useless.

Directly aft on deck, I saw a spar that I had tossed down in the clearing of the wreckage. I found that I could barely touch it and, by rolling it toward me, soon had it in my hands. Maneuvering its end into the open space, I found that I had the finest lever I could devise. It was about ten feet too long, but that didn't matter since it weighed very little down here, and there was ample room above.

I squirmed it around to fit a fulcrum and then dropped down to the deck below for maximum leverage. It worked! The anchor shank slid along, and soon all was clear of the companionway.

I squeezed out through the opening, stood on the main deck, checked my lines, and gave the three-pull signal, "pull me up" (with full decompression).

An hour later, I was climbing up the ladder of the *Silver Heels*. They explained what had happened. The engine had stopped, and their first thought had been to pull me up. Not being able to get an answer and not being able to pull me in on the lines, they decided that there was only one solution—a high-speed repair job. Uncle Hal found that the carburetor jet was clogged. He took it apart, cleaned it, and assembled it in record time.

We had a fine engine, but it wouldn't run on water or slime—and the gas that we were getting at Cuttyhunk was mostly those two noncombustible ingredients.

23

CAPTAIN CRAIG ARRIVES

We were just about licked on the *John Dwight*. We had been out there so long, it was hard to quit; we were so in the habit of visiting that wreck that we almost hated to leave. However, any day now, we knew, we would shove off for new horizons.

On one of those days, we returned to Cuttyhunk and found a telephone message awaiting us, "Max Gene Nohl, call Operator 12, New York."

I didn't know they had a telephone on the island, and when I saw it I still wasn't sure. There was a crank to patiently turn until a faint voice sounded and asked what I wanted.

There were only three things that I understood in the subsequent telephone conversation—it was Captain John D. Craig calling (was I familiar with his name?), there was an address where I could reach him, and could he come down and visit us?

Was I familiar with Captain John D. Craig? He was the Hollywood motion picture producer who had recently produced the picture *Sea Killers*, a story of a diving expedition off the Mexican Pacific coast; a lecturer of national repute; and a man who had experienced almost a storybook life of adventure.

He probably had gleaned little more from the conversation than I had, and so I immediately wrote him a letter trying to reconstruct our garbled chat. He certainly was welcome to come down and use our equipment and services in any way that he could.

A few days later, in came Craig, by the little ship operating once daily between New Bedford and Cuttyhunk. With him were two of his strange gang, each with a wild gleam in his eyes: René Dussaq, a native of Argentina

175

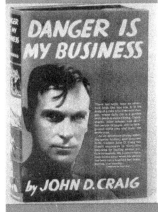

Nohl (left) and John Craig (right) diving on the sunken *John Dwight* from Nohl's salvage sloop off the Massachusetts coast.

who had also long made adventure his profession, and Tommy Larkin of Dayton, Ohio, a professional world traveler and cinematographer.

Craig was on lecture tour and had just a few days to stay, but we gave him ample opportunity to dive on the *Dwight* and to shoot some motion picture sequences both below and topsides.

Johnny was genuine. There was something about that man that I thoroughly liked. As he left, I strongly sensed that again our paths would cross.

There was one incident that happened (although hardly noticed at the time) that was to turn the entire course of my life.

We were talking one night, as we lay in Cuttyhunk after a day on the *Dwight*, about some of the fabulous treasure that had been lost at sea, and we were discussing the available depths to which a man in a diving suit could go to reach them.

Johnny was well familiar with the statistics relating to the world's record—the classical dive of the United States Navy diver Frank Crilley to a depth of 306 feet in the year 1915. We talked of the *Lusitania*, which we both knew was deeper than that. Johnny had apparently been thinking, too, in this direction. In my mind was going a very intense debate as to how much I should tell him. Finally, I could stand it no longer, and I broke down. "C'mon, Johnny," I said, "let's walk down to the boat—I want to show you something."

We ambled to the *Silver Heels* reposing serenely at her dock and jumped aboard. I opened up one of the under-bunk lockers, and there lying like an elaborately primped body resting in state, was a diving suit bristling with highly polished brass and chromium plate.

This was no conventional diving suit but the latest model of my long-dreamed-about brainchild—a self-contained suit. It had no air hoses whatsoever, the necessary gases for as much as eight hours' breathing being carried in the sparkling silver cylinders on the back.

Here was a diving suit that would unquestionably eliminate the many dangers of this hazardous profession, for it was in the air hose or compressor equipment that practically all of the fatal failures had occurred. No longer would it be possible for a diver to be fouled by his life-giving air hose. No longer would his air supply depend upon such complicated machinery as an air compressor and its driving gasoline engine. No longer would a diver be subjected to the terrific current drag of his hose in a tide. Here was a suit that would not require a half-ton of elaborate deck installations in compressor-tank-engine units but would be ready for diving by just getting into it.

Here—and now I was looking ahead—was a suit that would allow an experimental mixture of other gases besides air. Perhaps there was some other combination that might eliminate the dreaded diver's disease, the bends. For many years, I had been studying the physical characteristics of all known gases, as well as the known nature of that malady, and I thought that I had discovered the answer but still had to prove it in practice.

Johnny's eyes bulged.

"Someday, Johnny, we'll be able to go down four hundred feet in this suit."

Those words were to ring in my mind for many years to come. I repeated: "Someday, Johnny, we'll be able to go down four hundred feet in this suit!"

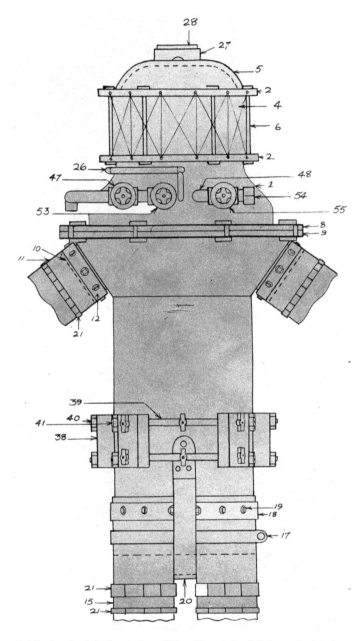

Nohl's sketch of the front view of his self-contained diving suit, dated January 3, 1935.

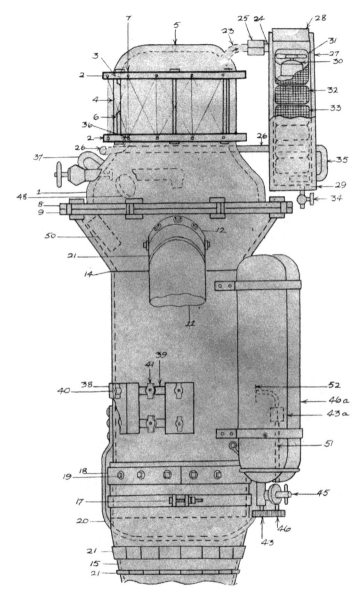

Nohl's side view sketch of his self-contained diving suit shows the two gas cylinders that made breathing possible without an air hose.

Book IV

The Bends in the Road

24

A New Venture

There was one thing that was definitive—we were done with the *Dwight*. We had gone back and made a thorough job of exploring her for her money, checking sample bottles of every section of her hold, and salvaging everything of value that was practical to remove. There was still a quarter of a million dollars in cash down there, but it apparently was not our lot to find it. We generously told our friends as we left that they could have the $225,000 with the steamship thrown in.

The *Silver Heels* was "up on the hill," parked high and dry in a Brooklyn boatyard. I was back home, in Milwaukee, wondering what was going to happen next.

One day, while I was wondering, the telephone rang. "Are you there, Gene?" asked a breathless voice.

"Hell yes," I answered, suspecting it was Johnny Craig. Most people in Milwaukee called me Max.

It was John. "I'll be right over," he said excitedly. "Gotta see you right away."

In a split second, there was Craig. "Remember that diving suit you showed me on the *Silver Heels*?" he asked.

I remembered.

"Didn't you say that someday that suit would go down into four hundred feet of water?" he asked, his gray eyes searching me inquisitively.

"I said that I thought so," I replied, "although there is still a lot of experimental work and research to be done before it will be ready to try."

"How would you like to go ahead with it," he asked, "right away?"

Craig addressing an adventurer's club with his lecture "Finding the *Lusitania*." COURTESY OF KATHY END

There was nothing that I would rather do, but what was the occasion of all this?

Johnny had been on a lecture tour in England. During the course of his public appearances, he had run across a group of people who were interested in salvaging the *Lusitania*. When I heard that name, I didn't care what else he said—whatever he had on his mind was a deal.

A man by the name of H. J. Demetriades, a Greek turned Scotch, had organized a Glasgow corporation to conduct salvage operations on the far-famed Cunard liner, lying someplace off Old Head of Kinsale, Ireland, in the open Atlantic.

Back in 1915, the United States Navy diver Frank Crilley had plunged to a depth of 306 feet to attempt to contact the sunken submarine *F-4* but had only been able to retain consciousness for two minutes under that tremendous pressure. In those two minutes, he had established a world record for deep-sea diving that had stood ever since.

The *Lusitania* was lying in 312 feet of water, one solitary fathom deeper than Frank Crilley's famous dive.

It was a lot more, however, than being able to go just six feet deeper. Sir

Malcolm Campbell might be able to drive his mammoth automobile, in a numerically similar analogy, at three hundred miles per hour over the salt flats of Utah or the tide-smoothed sands of Daytona to make a world land record, but that does not mean that the casual tourist can count on that as a practical rate for a Sunday afternoon ride. It would be impractical to allow, let us say, an hour for the three-hundred-mile run between Chicago and Detroit. A driver might consider sixty or seventy miles an hour as a practical country speed just as a diver might consider sixty or seventy feet as a practical working depth.

Thus, we had a long way to go to attempt such a stupendous salvage operation. But—there was, we understood, a $20 million treasure awaiting him who should succeed.

Before Johnny had said anymore, I announced that I was interested.

I asked him what sort of a deal Demetriades was willing to give us if we could devise equipment to do the job.

Johnny smiled all over, as is so characteristic of him, his eyes twinkling his overabundant enthusiasm. "The contracts are all signed," he said. "I thought you would be interested!"

Craig and I would go into a partnership on it—we would own together all of the physical equipment that was built; I would retain all my own patent rights in the event of dissolution; and I would (without salary) design, conduct all experimental work on, construct, and test all of the equipment. It was a loose sort of an arrangement, but Johnny was the one man with whom I would want to work on such a deal.

He was just starting a heavily booked winter lecture tour in this country, and I would not see him again for many months.

Practically at my fingertips was my good friend Jack Browne. Jack had been eating and sleeping diving for all of these years, too. He had not had a chance to actually do much of it thus far, except in the original five-gallon paint can helmet and during intervening summers when we had worked together in Lake Michigan with the professional suits that I had brought home from the East.

Was he interested? That was as foolish a question as I could ask.

In all of the dives that I had made with the original experimental suits, I was convinced of one thing—that the final suit would have to be more than just an adaptation of the conventional type of diving suit. I had been doing it this way for years but, over and over again, learned our major difficulties were not with the adaptation but in the fundamental design of the suit itself.

Thus, the plan was as follows:

First, we would build a breathing unit that could be strapped on the back of any diving suit, which would supply the diver with his oxygen requirements and purify his exhaled air. The design of this would be the result of everything that we had learned in many experimental units that we had constructed before. This could be tested by attaching it to our standard navy-type suits.

Secondly, we would build a super diving suit to go with the above unit.

Finally, we would have the opportunity to test artificial gas mixtures in the above apparatus.

We turned to and, in a few months, had our shop set up and fully equipped and had the new air units ready for testing.

There were a number of ifs, ands, and buts coming in over the cables from England, and Johnny was worrying about the stability of our venture and the value of our contracts with the Scottish corporation. Sensing that, we held off on any major expenditures, the new unit having been built up largely from parts of discarded former apparatus. Jack, always impatient at the slightest delay, decided to go ahead on a project that we had long talked of, the salvage of the SS *Westmoreland*.

I had long had this vessel listed in my files, since she carried down into the deep waters of Lake Michigan quite a substantial cargo of whiskey and gold—and what two things were traditionally of more lure to man?

There was one little detail that had caused me to record her as of only historic interest, and that was the fact that she had sunk in 1854, before the time of the Civil War.

The gold, of course, would not deteriorate. The whiskey was in oaken barrels, and it was undoubtedly true that the wood itself would not rot in the icy deep fresh waters of the northern lake. Jack had been in touch with many wood experts, all of whom agreed that the latter might be possible.

Cold, stark reality never fazed Jack very much—there was a small chance of making something on the wreck—and so I let him have one of the diving suits complete with hoses, pump, and telephones. He left, and I went back to work on the air-conditioning unit.

Craig (left) and Nohl (right) prepare for a test of a self-contained adaptation of the standard US Navy suit. Helium tanks are at their feet.

Every day or two, a letter would come from him. He was based over the eastern shore of the lake near Frankfort, Michigan. He had a boat. She was now fitted out with drags, diving ladder, air compressor, and everything that was necessary. There were both some pretty scenery and some pretty girls over there.

In Milwaukee, the self-contained unit was just about ready for a test. I was beginning to wonder about some of the forthcoming expenses—word from England was not good, and Johnny was apparently very hesitant about investing much more money in the new diving suit. Chartering a boat to test it would run about twenty-five dollars a day and would shoot the expense account up very rapidly. It was now at the point where I had decided not to ask Craig for more cash but to proceed myself on my own resources with the thought that in the event of the folding of the *Lusitania* expedition, I would then have this equipment as my own with a clearer conscience.

25

Human Guinea Pigs

Many months had slipped by but with no encouraging word from England. The *Lusitania* enterprise was suffering with internal difficulties—financial, political, and administrative. In the meantime, I was building up a few more underwater cameras, sweating over the drawing board on the proposed Super Suit, and getting everything in order for whatever might happen next.

And then one day, out of a clear blue sky, in blew Johnny, breathless, and with enthusiasm almost bubbling out of his ears. A cable had just come from England. We would leave in a few months.

"How soon can you build the new suit?" he asked.

I didn't know, but we would start in the morning. I showed Johnny some of the drawings and went over the theory of its operation with him. He was very pleased.

"Everything that can be chromium plated must be chromium plated," he requested, justifiably, since he was backing the bills.

Jack came back from Michigan, and we started to dig in. There were many problems to iron out, but in a few months there began to emerge a modernistic monster glistening with chrome and looking as though she could really do the job. The sight of that shiny suit was just too much, and one day Jack let the word slip out that the forthcoming invader of the *Lusitania* was waiting in our shop. In the twinkle of an eye, the place was overrun with reporters and photographers from the *Milwaukee Journal* and the cat was irretrievably out of the proverbial bag. The story went out on the AP wires and started a lot of things, good and bad, rolling. This would now, by law, be the announcement date, one year from which I

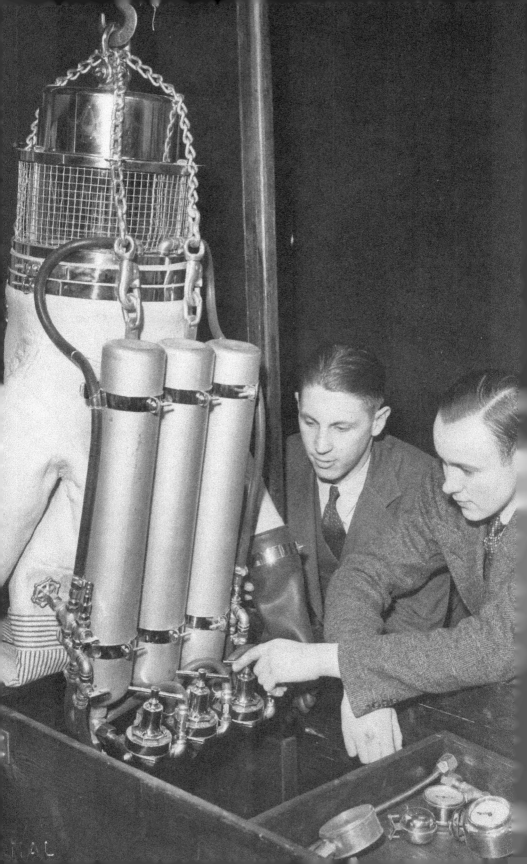

needed to file full patent applications. Hundreds of letters from would-be expedition applicants, question askers, and self-styled salvage experts started pouring in.

Among those many pieces of mail, there was one that immediately attracted my attention—from "E. M. End, M.D., Marquette University School of Medicine."

Dr. Edgar End had been much interested in the suit and our plans for using it, but there was one thing that had particularly caught his eye. That was the mention that had been made in one of the news stories that we believed we now had a practical means of administering an artificial gas mixture to the diver and that we had long been preparing to experiment with an oxygen-helium air.

By sheer coincidence, right in my own city of Milwaukee at the Marquette Medical School, Dr. End had arrived at similar conclusions as to the theoretical advantages of that gas in high-pressure physiology. Would I like to drop over and discuss our mutual interests?

I envisioned Dr. End as an old bewhiskered bifocaled medical research scientist with egg on his vest and formaldehyde reeking from his clothes. I had quite a pleasing surprise in store for me.

He was a young man—apparently just my own age—with a gracious personality so captivating that I felt the instant I met him that we would be friends for a long time to come. He was a member of the teaching staff of the medical school but devoted a large part of his time to research.

Curiously enough, Dr. End had not developed his interest in helium from deep-sea diving but had approached it from an entirely different direction. Working with Dr. Percy Swindle, the eminent head of Marquette's Department of Physiology, he had developed a deep interest in the agglutination of red blood cells. Going into the many manifestations of this physiological phenomenon, he had planned to expose animals to various gases under varying pressure conditions.

Helium had under pressure resulted in a decided effect on agglutination, and then the question had arisen: what has agglutination to do with the bends?

FACING: Nohl (left) and Browne (right) with prototype II of the self-contained diving suit.
COURTESY OF KATHY END

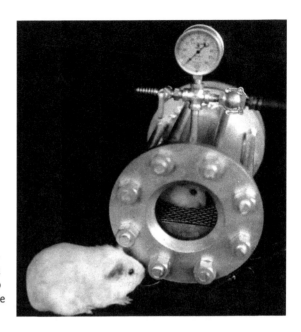

The high-pressure experimental animal chamber at Marquette University. Two guinea pigs prepare to take a dive.

Dr. End had built a small chamber in which he could put mice and subject them to varied pressures to study this phenomenon.

He had arrived at helium for diving from the back door, and I had arrived at it from the front door from entirely different considerations—but here we were, in the same house with the same objectives in mind.

We shook hands on it and were off. Dr. End offered me laboratory space that I could have for my own, and in a few days, we moved over bag and baggage, tools and diving suits, and started digging in.

First, we would try our proposed helium air on animals. Dr. End's first tank for mice was not overly spacious or overly deluxe. He had rigged it up out of an old piece of two-inch waterpipe, thus limiting diving service to a maximum of two small mice. We decided to design and build a larger one with accommodations for several full-sized guinea pigs capable of withstanding terrific pressures. It would be fitted with a large pressure-withstanding window through which we could watch them as they took their "dives."

In the meantime, I finished up the final touches on the new suit, and it was now ready to try. There were quite a number of questions on weight distribution, negative buoyancy, possible leakage, air control, etc., that I

The first dive with the new suit in the pool of the Milwaukee Athletic Club. Jack Browne tightens the collar band, while Nohl is in the suit. *MILWAUKEE JOURNAL* PHOTO

could test just as well in seven feet of water as in two hundred. Thus, we decided to take the first dip in the treacherous depths of the Milwaukee Athletic Club swimming pool, where we could find a full two fathoms at its deep end.

The apparatus worked like a charm. I wanted to spend a few hours underwater to see how the chemicals reacted to sustained exposure and

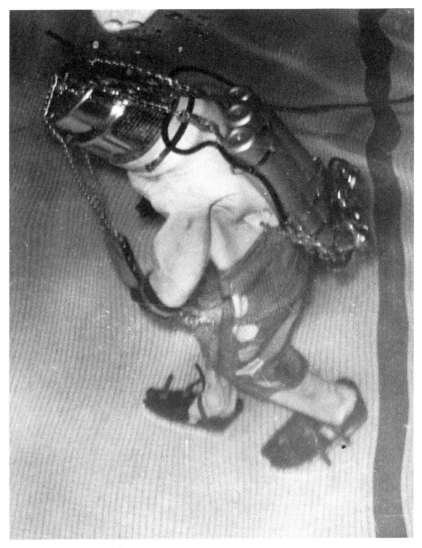

Two fathoms deep. Nohl walks along the bottom of the Milwaukee Athletic Club pool in the first test of the new suit. *MILWAUKEE JOURNAL* PHOTO

to run a check on the amount of oxygen I was consuming, so I just sat down on the bottom and leaned against the tile end wall. Chancing to glance upward, I could see scores of curious shimmering faces staring down through the clear water, the entire end of the tank being lined up with curiosity seekers.

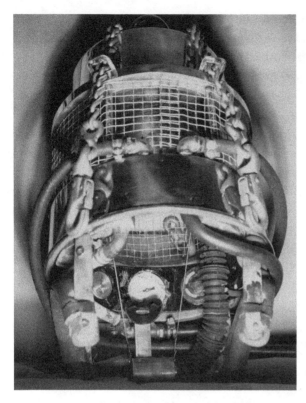

Instruments within the helmet include a watch, oxygen gauge, mouthpiece, microphone, and earphone. COURTESY OF KATHY END

I felt very self-conscious. Where *could* a fellow go to find privacy?

For three hours, I amused myself by singing. There was not even a fish to complain.

We now had a supermodernistic diving suit that was ready to go to work. It was a suit with many improvements over the standard conventional equipment. There was no air hose. That immediately freed the diver from the dangers of compressor breakdowns. There was no danger of fouling, and we could now venture into wreckage that would be unthinkable to penetrate with an air hose trailing behind. There would be no drag when the tide currents ran. The suit was so simple that a man could get into it in a minute's time, whereas it was usually an hour's job to dress a diver in standard gear. There were many convenient built-in gadgets: a radium-faced clock, oxygen gauge, compass, microphone, and telephone speaker.

These things were all very nice—but fundamentally, all of the above were not going to take us down to the *Lusitania*. We still were breathing air and were subject to all of the limitations of air in depth. The real important thing that we now had was a diving suit that would allow us to breathe any possible combination of mixed gases that we should care to try.

A short explanation of the diver's disease, the bends, might here be appropriate.

The classical explanation of the bends has always been the formation of bubbles throughout the fluid tissues of the bodies as pressure is too rapidly decreased. This is a phenomenon familiar to everyone in the common analogy of the soda water bottle. At the bottling company, a gas—carbon dioxide—is dissolved in the beverage under high pressure. The pop soaks up and retains large amounts of it in solution as long as the bottle cap keeps the pressure in. However, when the cap is removed, the pressure immediately drops and the soda water can no longer hold the excessive gas in solution and out it goes in the form of bubbles.

The diver, the theory explains, is nothing more than a bottle of pop, the high-pressure air that he is breathing similarly dissolving throughout the blood and tissues of his body. When he is dragged up to the surface, the pressure falls off and he can no longer hold the excessive quantities of gas in solution.

He fizzes!

From this same analogy, several more conclusions may be drawn. If the pop bottle, instead of being opened suddenly, is merely punctured with a pin prick through the metal cap, there will be no fizzing. The pressure falls slowly and allows the gas to emerge from the solution in an orderly retreat, without the formation of bubbles. Likewise, a diver brought up slowly to the surface passes off the excesses of gas into his lungs from the bloodstream with no trace of the bends. This, then, is the basic theory of decompression.

If that fizzing bottle of soda pop can be put under pressure again, the bubbles will immediately decrease in size and redissolve into the liquid— and soon they will disappear. This in a diver is called recompression and is the only treatment for the bends that has ever been discovered.

It would seem as if the logical attack would be to look for a gas that was physiologically inert and that also had a very low solubility in blood in comparison with air.

A number of gases appeared as possible prospects: xenon, argon, krypton, neon, and helium in the inert gas group. There was also hydrogen. Of all of these, helium stood head and shoulders above the rest as the answer—it was safe, plentiful, and cheap.

About seventy years ago, the French astronomer Jules Janssen discovered, through his spectroscope, ninety-three million miles away in the fiery tongues of the vapor on the sun, the presence of a new element as yet unknown to man on the planet Earth. It was named helium from the Greek *helios*, the sun.

Forty years later, it was found to exist right here on earth, mixed with the air of every breath that we take, but in such infinitesimal quantities that its cost of extraction—retailing for ten thousand dollars per cubic foot—kept it strictly in the class of a laboratory curiosity.

A large reserve of helium was discovered at Dexter, Kansas, in 1903. It was believed at first to be a rich well of natural gas, and the entire countryside turned out to usher it in with a huge celebration. The mayor of Dexter was to light a huge tongue of flame with a torch as a spectacular climax to the ceremonies—but the gushing gas put out the torch. The natural gas that wouldn't burn put Dexter down in history. It was helium.

The lightest element in the world is hydrogen, a highly inflammable, highly explosive gas that in combination with oxygen forms water. The next lightest is helium, a totally unburnable and practically chemically inert element.

Although helium is almost exactly twice as heavy as hydrogen, the efforts to use the latter in lighter-than-air craft have almost inevitably met disaster. The dramatic burning of the giant dirigible the *Hindenburg* in 1937 practically pronounced the death knell for the use of hydrogen for such purposes.

Because of the low density of helium, it appeared to us that another factor might contribute to its effectiveness for divers—diffusibility. This gas would go out of solution much more readily than the sluggish nitrogen, allowing a diver to rid himself of it in less time—thereby shortening decompression.

There was another factor that now suddenly tipped the scales in favor of helium.

Every deep-sea diver is familiar with the intoxicating effects of high-pressure air. At 100 feet, there is a groggy, disorganized feeling that is much like a partial alcoholic inebriation. At 150 feet, it is difficult to stay conscious. At 200 feet, only a rare diver can retain his senses. At 306 feet, only one diver had ever been able to experience the sensation—Frank Crilley—and he passed out in two minutes.

This, always, in all of the books on diving, had been called "oxygen intoxication." There never had been an explanation as to where it had assumed this name other than, "What else could it be?" Nitrogen is inert, and what else is there in the air besides oxygen? (Air is approximately one-fifth [21 percent] oxygen and four-fifths [79 percent] nitrogen.)

Somehow or other, it didn't seem to make sense. We had been using pure oxygen on a lot of dives in our new suit—and instead of experiencing the supposed intoxication, our heads were clearer than usual.

Could it be that the nitrogen was causing the trouble? That nitrogen under pressure was no longer an inert gas? That the replacement of nitrogen with a truly inert gas, helium, would lick this problem as well?

There was one way to find out—to try it!

First, however, we would try it on the little guinea pigs.

We did. We could bring them out of the water in less than the normal decompression time with the helium air, and yet, with real air, just like their distant cousins the human beings, they came down with crippling cases of the bends. Their minds seemed clearer under great pressures. Their orientation was decidedly improved.

Many, many years ago, at the time that the crib was laid off the shore of Milwaukee, there had been an accident. A number of men working in the caisson on the deep floor of Lake Michigan were killed by the bends.

To many horrified Milwaukeeans, this was only a dramatic but unfortunate accident. To one man, this was an accident that, if he could help it, would never again happen. Joseph Fischer, for many years the chief engineer of Milwaukee County Institutions, instigated a crusade to see that it would not. He well knew that had there been a pressure chamber into which those men could have gone to recompress, many or all of their lives could have been saved. Such a pressure chamber would cost thousands of dollars to build and might never be used; but if it should someday save one life, it would be well worth its cost.

Through his tireless efforts and from his design, the chamber was built, and it was installed in the basement of the Milwaukee County Hospital.

It was the only pressure chamber in the United States inland from the Atlantic or Pacific coasts. Now for many years, it had stood there, still resplendent, unused, in its shiny coat of silver paint and almost forgotten except for a few prowlers in the hospital basement.

Dr. End knew of it. My ears pricked up—this was just the thing he wanted.

Could we use it? Joe Fischer veritably beamed. Why, he would drop everything the minute we were ready and personally supervise the entire operation of the huge tank.

We loaded up an automobile with spirometers, canisters, huge cylinders of oxygen and helium, valves, mouthpieces, breathing tubes, and a lot of Dr. End's equipment for various physical observations as we exposed ourselves to pressure.

We were keeping Johnny Craig posted of our progress. As we were about to start on the experiments with the new synthetic helium air, he wired us and said that he would like to be one of the "guinea pigs." That was fine, and so we postponed the first set of dives until a time when he had a few open days on his lecture tour.

Johnny and I were going to breathe the helium air from the apparatus, and Dr. End was going in with us without it. We would all go under pressure together, remain under pressure for a full hour, but then, decompress separately. We, with helium in our bodies, would try to duplicate what the little guinea pigs had done. Dr. End, with air in his body, like a diver, would have to decompress separately in the back room of the tank.

We realized that we had no right to assume that just because those fuzzy little guinea pigs had successfully dived with the new gas, that we could do it, too. We decided to do this in the intelligent way. On the first day, we mixed two parts nitrogen with one part helium and cut our decompression time proportionally.

It worked! We felt fine! It seemed as if our minds were considerably clearer than they would be at that depth (one hundred feet) of seawater, breathing pure air.

The next day, we diluted the nitrogen with two-thirds helium. It worked again. Our heads were remarkably clear. The third day, it was a three-thirds dilution with helium—in other words, there was no nitrogen at all! This was then a mixture of 21 percent oxygen and 79 percent helium. The decompression was to be rapid. This would be the dive that would tell the story.

That tank was no place for sufferers from claustrophobia. As soon as Dr. End slammed the one-ton door shut, there was an immediate terrifying sensation of oppression. The reverberation of the slightest noise sounded like a landslide. The roar of the air coming in from the compressor sputtered like a machine gun at our ears.

There was a terrific clanging as someone was tightening something on the outside with a wrench. The pressure was now mounting. I felt a dull pain in my head. A little combined yawn-swallow and my tubes blew, and everything was clear.

The heat was getting oppressive. Our sweating bodies were saturating the tank with humidity and the odor of perspiration. The heat of the compression had now turned the tank into an oven. We just sat there and stared at the spirometers bobbing up and down as we exhaled and inhaled.

About ten minutes later, the telephone clanged, sounding like a fire alarm in the air-packed space. Dr. End answered it, and we heard Joe Fischer's voice sing out, "Forty-four pounds," resounding through the tank wall more audibly than it did through the phones.

"OK," confirmed End.

A moment later, the compressor was turned off, leaving us in a contrasting silence that was almost alarming. For an hour, we sat there and amused ourselves by just breathing, staring straight ahead at the spirometers. We couldn't talk, since we had the breathing tubes in our mouths, but we had ample opportunity to think. I seemed to sense a mental clarity that I had never known before at this depth. I recalled the many scores of hours that I had spent on the *John Dwight* at about this pressure and the inevitable sensation of intoxication I had always associated with that rum runner without as much as tasting her rum. I had simply been drunk on nitrogen. Dr. End was a busy bumblebee, checking blood pressures, temperatures, and pulses and listening to our hearts through his stethoscope.

An hour later, the phone clanged again, and we heard Fischer officially announce that our exposure time was up. End was now to leave us and seal himself in the back room. He said "so long" to our unanswering forms as we rolled our eyes to see him depart through the massive back

Chief engineer Joseph Fischer at the controls on the exterior of the recompression chamber. Pressure is at forty-four pounds per square inch, equivalent to one hundred feet of seawater. *MILWAUKEE JOURNAL* PHOTO

door of our chamber. There was a lot of clanking as he sealed himself in. We then heard two knocks through the tank wall to start our rapid decompression.

Suddenly, there was a shrill scream as the air outlet valves were cracked and swung wide open. The air was rushing out at hurricane speed. Immediately, we felt our lungs expanding and saw the spirometers rise higher and higher to automatically valve huge globs of the expanding helium air. We grabbed frantically for the blankets hanging over the backs of our chairs as the temperature had plunged precipitously to what was probably about freezing from the expansion of the air. Our wet open-pored bodies shook with cold as we sensed that same sensation that every diver knows of rapidly falling pressure.

A few minutes later, the massive steel door swung open, and it was all over. Those valves had passed over two hundred cubic feet of air in that time to get us back down to atmospheric pressure.

There were all sorts of people standing around outside—nurses,

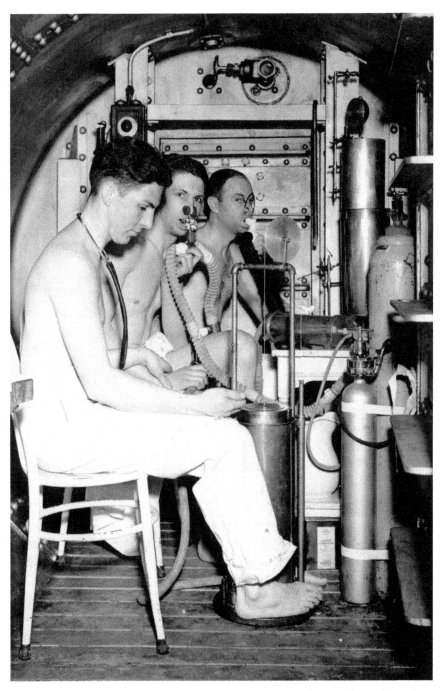

Inside the pressure chamber. Dr. End at left, checking Nohl's pulse. At right is John Craig.

All over. No bad symptoms. Left to right: Nohl, End, and Craig. COURTESY OF KATHY END

doctors, photographers, reporters. This must be quite an occasion, and so we dug up our old diving gag.

"What's that funny smell?" I asked Johnny.

"Fresh air!" he answered, as if he had never said this before.

Now, after a half a dozen years of dreaming and hoping, I had actually seen it work. Not only had it worked—the results were amazing. We had no bends—not even the slightest trace of that familiar creaking pain that I had so often felt with full specified decompression. Even more important, under that full pressure, which had been a weight of approximately 132,000 pounds over our respective body surfaces, neither Johnny nor I had felt the slightest trace of grogginess.

Nitrogen, apparently, was the culprit. Helium would, apparently, be the answer.

26

GNARLED ON THE NETWORKS

From now on, it was just routine.

The suit worked like a charm with plain air. The helium worked like a charm in the tank. We had gone as deep with the suit as the air would permit. We had poured as much pressure into the tank as its one-inch-thick steel wall would permit under safety regulations.

All that was left was to combine the two and start going down deep.

I chartered a boat and soon found myself wading around in a hundred feet of ice water far out in Lake Michigan. This was a mere combination of the above elements, and it worked out as expected.

Johnny blew into town again. He was bubbling with enthusiasm, and I knew something was up.

"How'd you like to do a radio broadcast?" he asked.

The thought of a microphone ran a chill down my back. I had done a lot of sound effects on the *Seth Parker*, but what was this?

"Our own NBC show," Johnny went on to explain. Ken Fry, director of special events for the National Broadcasting Company at Chicago, had asked if we could put on an underwater broadcast with the new suit. I couldn't think of any good reason why we couldn't.

It would be a Saturday afternoon show. Johnny had the weekend off on his lecture tour and would fly in to Milwaukee. Jack and I would get everything ready.

WTMJ, the *Milwaukee Journal* station, was so interested that they wanted not only to fee the NBC network but expressed a desire to take an additional half hour beforehand. This would make a continuous hour that we would broadcast.

Nohl, preparing for a test (in the Milwaukee River) of "Old Droopy Drawers," the conversion of navy apparatus to a self-contained diving suit

It was not practical to have more than a three-way hookup—I was to wear the new helium suit, Johnny to wear the regulation navy suit, and Russ Winnie, WTMJ's genial announcer, to act as master of ceremonies for the entire hour.

Jack vigorously resented the plan. He had been working hard to help me, but there were not enough diving suits now with electrical hookups available for him to be on the bottom, too. He didn't hide his sentiments. Johnny (who only blew in on such state occasions) was basking in the lime-light, and Jack (who with me did all the work) was left merely to talk to us up on a desk. Accordingly, we decided to work out a combination that in the end was to almost kill us all.

Jack would go down in the navy suit before the broadcast started. We would open the show on WTMJ with him on the bottom, and I would be starting down into the water. At the fifteen-minute interval, I would land on the bottom and Jack would arrive at the surface to start undressing. Johnny would then get into the suit, being previously partly dressed in another suit, so as to be ready to go overboard at the end of the half hour. At that time, we would transfer to the NBC network. Johnny would drop down and meet me below, and we would spend the last half hour together on the bottom.

To add color to the locale, we decided to dive on the wreck of the sunken steamship the *Norlond*, a freighter lying in eleven fathoms (sixty-five feet) of water off of South Milwaukee in Lake Michigan. She had carried quite a substantial cargo of white lead, which might have been good salvage if the ship had not been too badly broken up.

We had arranged to use the Coast Guard cutter *Antietam* for the broad-cast, shooting the show to shore by shortwave to the WTMJ station, which would then feed its own and the NBC wires.

Jack was still very depressed that he could only be on WTMJ and not on NBC, so much so that Johnny was anxious to leave him out altogether. It was too late now to make the change, however, and so we went ahead with our original plan, the rush of which resulted in the subsequent difficulties. The big moment came.

Jack went down on schedule, soon after which we were on the air. Fifteen minutes later, I was ready to go overboard. Getting down into the water, there were so many things to think about that I hardly remembered we were broadcasting. Russ Winnie had a smooth, soothing line of patter, describing vividly everything that was going on and keeping up a running conversation with Jack on the way up and with me on the way down. As the equipment changed from Jack to Johnny, Russ and I gossiped as best we could.

Conversation from a diving suit is not like talking in a sound-proofed studio. The self-contained suit had a decided advantage over the navy suit

Nohl takes a test dive from the *Antietam*, April 10, 1937. COURTESY OF KATHY END

in that the latter was almost smothered under the terrific roar of the air compressor, whereas in the new apparatus, there was just an almost indistinguishable hiss of the replenishment oxygen. However, the mere process of breathing kicked up quite a roar.

Talking in compressed air is as strange a sensation as any man can experience. The words seem deep and throaty and are hard to distinguish. The reverberation in the helmet space is overpowering. Because of the change in density of the molecules, it is utterly impossible for a diver to pronounce the letter s—a thort of lithp rethulting.

"A few fith are thwimming thwoo the weckage," I went on to describe. "I'm crawling over thum thlimey timberth. I can thee about thwee or faw feet."

All of this took quite a bit of time, since each phrase had to be repeated and then translated by Russ. "A few what are what?" he would say.

"A few fith are thwimming," I would repeat, trying as hard as I could not to lisp, but God himself couldn't sound a single s submerged in this watery world.

"A few what?" he'd re-ask. "Oh, a few *fish* are what? Oh, are *swimming*," he chuckled.

Thus went the first half hour.

That morning, the question had come up, "Should we plan to have some little dramatic incident—some little impending accident to add that certain element of suspense?"

"Hell no!" was my response.

"Hell no!" was Johnny's response.

The minute something like that started to happen, the typical critical radio listener would inject a mental "Oh yeah!" That such a thing would happen at just this time, with the entire facilities of the National Broadcasting Company listening in, would generate the faint stench of phoniness.

"Whatever happens down there, let's not overdramatize it—we'll just make it a straight dive, even if we are actually in the process of being eaten by an octopus," we emphatically agreed.

This, I still believe, was good showmanship. But as the proverb states, sometimes, the truth is actually far stranger than fiction.

Everyone was in a dither. Jack had overstayed his allotted period and didn't give Johnny time to get into his suit to go overboard on schedule. I was still down there on the bottom, prowling around as they were racing Johnny into the gear.

I heard Russ start out, "Ladies and gentlemen, it is our pleasure to bring to you, over the facilities of the National Broadcasting Company, a unique program in the deep waters of icy Lake Michigan."

Russ called me and asked me to describe the wreckage, and then as we hashed over my lisping, he cut and announced that Craig was now going overboard.

"Are you ready, John?" he called.

"OK, lower away," answered Johnny.

Visibility being at half a fathom, I couldn't see Johnny, nor did I have the slightest idea of where he was, although we were carrying on a running conversation with each other and Russ.

Johnny's lisping, muffled voice sounded "terrific," dripping with atmosphere; once I heard it and realized that my own was just as bad, I knew that the broadcast was a success.

Johnny lit out like a scared rabbit looking for me so that we could enjoy the comfort of being within arm's length, even if we could not see each other. Suddenly, I felt something pull over my back. I turned and saw that it was his air hose.

"Your hothe jutht cwossed my back," I called up through Russ. He repeated the message and apparently Johnny was now trying to cut over toward me.

No one will ever know just exactly what did happen. The next thing I knew, the hose crossed again in front of me and then drew taut. I was in a coil. Further examination revealed that there, also, were his lifeline and telephone cable.

I thought that the best thing I could do would be to step out of the tangle. But it was too late.

That snare suddenly tightened, and I felt myself being yanked off the deck with a terrific jerk. It happened so suddenly that I had no time to grab anything. We were shooting up toward the surface at incredible speed.

I heard a dull thud through the water, and we stopped. I twisted my head in the 360-degree-vision helmet and saw, off my left shoulder, the keel of the Coast Guard cutter. We were under her bilge.

Russ called John, apparently totally oblivious of the fact that we were up at the surface under the bottom of the ship.

"Do you see Max?"

There was no answer.

He called again.

"John? Calling Johnny Craig? John, do you see Max?"

There was still five minutes to go before we were off the air. There would be no dramatics, we had decided.

I went on to describe the water, the milky haze, the funny feeling of being in a diving suit. We joked back and forth. Russ was a smooth talker

Craig, bloated like a balloon and unconscious, is towed in to the Coast Guard cutter *Antietam*. Russ Winnie, WTMJ announcer (at top with earphones and microphone), now sees why he has been unsuccessful in calling him. Nohl is still in the water.

and knew just what to say or ask on the spur of the moment to make it interesting.

And then I heard him cut in, "We regret, ladies and gentlemen, that our time is now up. We have brought you this broadcast from the deck of the sunken steamship, the *Norlond*. This is the National Broadcasting Company."

There was a pause of three seconds. And then Russ boomed in, "Max, we're off the air. What in the hell is wrong? Where are you?"

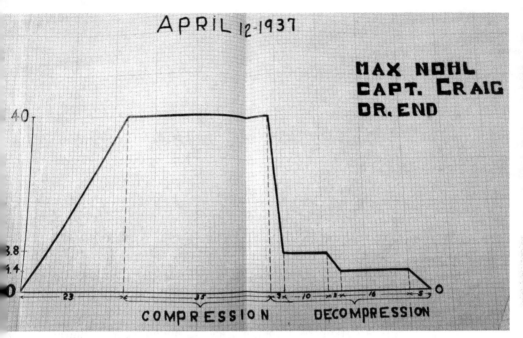

Decompression graph created on April 12, 1937, two days after Nohl's test dive on the *Norlond* in the first prototype of the self-contained diving suit.

It turned out that we were not completely off the air—we were off NBC and also WTMJ, but the shortwave station was still on for those who chose to hear it that way. "We're in an awful snarl just under the bilge of the *Antietam*, I think—John passed out," I explained, and in a little while, they pulled us on.

Johnny, who was now gradually gaining consciousness again, had soaked up a suit of woolen underwear with blood, partly from the bump and partly from his lungs, a form of the bends resulting from the rapid expansion of gases. I had been dangling a length below Johnny and had not bumped my head nor had I the slightest trace of the bends, although we certainly had gone up together under identical conditions. I had actually been down there half an hour longer than John. He had been breathing air, and I had been breathing helium. It worked!

27

THE ELUSIVE *LUSITANIA*

My next dive was 150 feet, far offshore in Lake Michigan. I felt fine. A week later, I hit two hundred feet. The helium was nothing less than magic. Although it was possible to go this deep with air, it would be a fight just to stay conscious; with the helium, my mind was clear as crystal.

We were now partially ready for the "Lucy." Although she was lying on a 312-foot bottom, a vessel of that size was, in proportion, like a rowboat in three feet of water. If she was on her side, as we had assumed from the accounts of her sinking, our shoes would strike her plating at just about two hundred feet. If her starboard side was up, as we much hoped, the strong room would be on her upper side.

We now had a diving suit that would reach this depth—a hoseless suit that would allow us to penetrate the interior of the wreckage without fear of fouling, a suit that would allow us to work without fear of losing consciousness or picking up the dreaded bends. We could continue our testing into deeper water "on the job" as we had need to go down into the hull interior.

Johnny was handling all of the contacts with England and Glasgow, and so we in Milwaukee concentrated on last-minute details preparatory to sailing. Jack, Dr. End, and I had our suitcases all packed, the suit crated up, and all patterns ready so that we could order more equipment as we needed it. Johnny wanted to wait until we had actually started on the job to build more of the not-inexpensive diving suits.

There was still one little niggling detail—the wreck had to be located on the floor of the Atlantic Ocean.

We had always passed this off as nothing more than a detail on assurance to that effect from over there. They had Captain Turner's chart on which her course had been computed up to the moment she sank. There, on that piece of paper, marked by a grim X, was her position at the time she went down.

Even if there should be any more of a problem than we had anticipated in picking her up at or near this theoretical X, it could not take long with the equipment we had to locate her.

The salvage vessel, the SS *Orphir*, of Glasgow, was fitted with an amazing device that would draw on a smoked drum the contour of the bottom over which the surface vessel was passing. This, the British Admiralty Supersonic Echo Sounder, made a picture of the bottom with a sound wave!

Sound travels well in water and at approximately four times the speed that it does in air. A sound wave radiated from a surface vessel will reflect off the bottom and return again to the ship just as the familiar air echo would reflect off of a large cliff or building. To time this round trip, knowing the exact speed of sound, obviously makes it possible to calculate the depth.

The apparatus did just that but entirely automatically, sending out and picking up the echoed signal at extremely short intervals and actually recording a scale contour of the bottom on a revolving smoked drum.

In passing over a massive hulk like the *Lusitania*, a decided change in contour would be immediately apparent. Although she was lying at a depth to which no diver had ever penetrated, her upturned side would reach one-third of the way to the surface.

Thus, locating the wreck was merely a matter of maneuvering over an area in which she might still be until the supersonic recorder picked her up.

At the time that she was struck by a Nazi torpedo on May 7, 1915, the *Lusitania* had been steering in a zigzag course, apparently having zagged when she was supposed to have zigged. At the end of the last zig on the chart was an X marking her grave. The corresponding spot on the ocean floor would logically be at the center of probability of her position, but realizing that this position was not fully determinate, Captain Henry Russell, master of the salvage vessel, laid out an area of approximately a square mile inside of which she must certainly rest.

Day after day, the *Orphir* patiently steamed back and forth in this theoretical plot, crossing and crisscrossing to cover every square inch of it with the incessantly bounding signal.

There was no *Lusitania*.

Perhaps she had coasted farther than Captain Turner had calculated before she plunged her huge hull into the cold green water of the North Atlantic. Captain Russell again patiently laid out another area of the next highest probability, dead ahead on her last ill-fated course. He buoyed it; and started to comb it with the supersonic ear. There was no *Lusitania*.

Perhaps she had not coasted as far as Captain Turner had estimated. Another square mile was laid off and the *Orphir* patiently and meticulously crisscrossed this area.

There was no *Lusitania*.

She might have been farther offshore than Captain Turner had realized. Another mile was laid off and surveyed. Failing again, the inshore possibility was covered to complete what resembled a huge cross. With no luck, the corners were next filled in to make a giant square comprising nine square miles.

There was no *Lusitania*.

Captain Russell was a veteran of the sea and nothing but praise could be paid to his thoroughness and tirelessness in this highly discouraging job. Operating the *Orphir*, a 175-foot steamship, with her crew of twenty-two men was a matter of enormous expense to the syndicate. Tens of thousands of dollars were disappearing into thin air like the wisps of steam from her exhaust. It is not difficult to understand that he was impatient during all of these months of the fast-waning summer at the countless ridiculous suggestions that were constantly coming to the attention of his overworried mind, suggestions largely from land-lubbing self-styled salvage experts who weren't quite sure as to which side was starboard and which was port.

There were electrical geniuses with monstrous, fearsome-looking machines. There were prospectors with equipment used for locating mines. There were men with divining rods, strange wild-eyed creatures who guaranteed that with a forked willow twig in their hands they could feel the pulsations of such a mighty hulk when they passed over it. There were psychic characters who claimed they would just know when they were over the wreck and who would need no electrical instruments or willow twigs. There were the fishermen, many of whom claimed to have some possible clues. Captain Russell was being besieged by these people. He was a busy man. He had a big job ahead of him. It is understandable that he didn't want to as much as talk to the assorted variety of wild-eyed screwballs.

He made his plan, a thoroughly scientific method of attacking the problem, and had, in his determined manner, decided to see it through, right or wrong.

It had probably been wrong, he was finally willing to admit. It was not that they had not covered thoroughly the allocated area but that Captain Turner had probably been in error as to the location of the ship at the time that she was struck. After all, she had been zigzagging, and it is almost impossible to chart an accurate dead-reckoning course under these conditions.

"Why don't you at least talk to the fishermen?" came forth from one of the boys in his own crew. "They know more about these waters than anybody."

That, perhaps, would be an excellent idea. The scientific approach obviously hadn't produced the *Lusitania*.

It was announced that on the following Friday night, there would be a party at a designated pub. Anyone who knew anything about the *Lusitania* was invited, and there would be all the beer that a man could drink.

Russell didn't think it would be much of a party, but he officially put in an appearance with the charts under his arm.

There were over forty people there, each of whom had seen with his own eyes the *Lusitania* sink. After all, she had gone down only a dozen miles off shore. Children had been let out of school to see her in her death throes.

Captain Russell hastily unrolled his charts of the waters off of Old Head of Kinsale. The grizzled fishermen, steins in hand, clustered around.

"I was standing right here," explained one of them, transferring his beer to the left hand as he pointed to a spot on the chart. "I lined her up with the edge of McGinty's barn, which is right here," he said, moving his weather-beaten finger to another point, "and this big rock, which is right here," he indicated, pointing to the third position.

Russell took his rule and made a line through them, extending a score of miles out over the water. It did not cross his carefully dragged area by several miles.

Another fisherman similarly pointed out two objects that he had lined up with the sinking ship from his vantage point on that dismal day for the world in 1915. Russell, again, drew in a line connecting the two and extending it far to seaward. It crossed the first line. If those two men were right, the *Lusitania* was lying at their intersection, a point about twelve miles offshore in fifty-two fathoms of water. A third observer came forward and gave his observation—and the seaward bearing line crossed the other two at almost the same point!

Before the evening was over, Russell had drawn forty lines from as many different shore locations extending to sea, and they all crossed at almost exactly the same spot!

The next morning, October 6, 1935, Captain Russell steamed out of Kinsale Harbor for this theoretical spot. Within twenty minutes, he was over the wreck.

The long lethargic stylus danced crazily as the *Orphir* passed directly over the one-time "Queen of the Seas." There, before his eyes, on the revolving smoked drum, was an almost fantastic photograph—a picture made not with light in the dim deep waters but with an echo of sound. There, as clearly as the most credulous could demand, was the mighty hulk of the great long-lost Cunard liner.

The image of the drum was like a silhouette. From the speed of the *Orphir* and the scale of the stylus movement, the dimensions of the wreck, apparently lying flat over on her port side, could now be checked.

As the *Orphir* passed over her, the height of the protrusion above the bottom should give her beam. It was almost exactly eighty-eight feet—that of the *Lusitania*. From this same silhouette, the distance from keel to her main deck could be measured from the scale of the speed of the *Orphir*. It was almost exactly sixty feet—that of the *Lusitania*.

It was she!

BOOK V

YOU HAVE TO HIT THE BOTTOM TO REACH THE TOP

28

Deep Water

They had located the *Lusitania*.

However, it was now October. The full fury of the North Atlantic had broken loose, and its mighty seas were sweeping over her long-hidden grave as if to display the scorn that Father Neptune flays before those who would rob his dead.

The summer had gone, and we had long known that no one would ever work in those unprotected waters as late as this. The *Orphir* had been tossed about like a chip in Captain Russell's last frantic attempts.

We, waiting and building equipment in this country, had long since realized that the summer was gone, although we had been ready to leave at a moment's notice. Dr. End aptly expressed it when he said, "I wore out three suits of clothes just packing and unpacking them." In the meantime, we were building new cameras and diving equipment.

The suit was ready for deep water, and I was itching to get down into it. Lapping up into our front yard at Fox Point, a suburb of Milwaukee, were the frothing seas of Lake Michigan. For too long, I had looked out at those waters and prepared to go down deeply into them—it was now time to do it. It would be next spring before we could use the *Orphir*, and so we looked around to see what we could get in floating equipment out of Milwaukee.

I had, in the meantime, hit three hundred feet and felt as fit as a fiddle last spring. Three hundred feet! That was "only one lousy little fathom"—just six feet—less than the world's record.

There were still a few more test dives to be made in the summer's improvements to the suit, but of most importance, there was the problem of getting down into the unknown. Studying the charts, I found that by heading northeast from Milwaukee toward a point about thirty miles out, we could hit almost four hundred feet of water, a depth not too easy to find in Lake Michigan.

I was anxious to have a good crew up on deck for this dip down to a depth that I didn't know much about. It would be December and bitter cold out there. We had to have facilities for keeping the boys up topsides warm. No bobbing little cabin cruiser would do for this job.

The United States Coast Guard had been good to us and had allowed us to do our broadcast last April from their cutter *Antietam*. Perhaps, through the same channels, I could get it again. I went down to see Ken Fry, director of special events in Chicago for the National Broadcasting Company. He had done most of the arranging for our broadcast from the SS *Norlond* when Johnny Craig had snarled himself up.

"Of course," said Ken, "I'll see what I can do—not only that but how about taking an NBC microphone down with you?"

That was a hard invitation to pass up. Ken fixed everything.

A few days later, a huge NBC truck came roaring up to Milwaukee, and we all piled in and drove north to the city of Port Washington. This would be the closest town to the chosen diving location. At the very crest of the characteristic hill of Port Washington was the Catholic church, with its massive steeple towering skyward. "What a place for an aerial!" cried the engineer as he spied the lofty structure. The broadcast, similar to the other, would be sent ship to shore by shortwave and then telephoned into Chicago to feed NBC through its key stations.

The *Antietam* had always looked enormous to me, perhaps in contrast to our own small boats. However, it didn't look so big on the morning of December 1, 1937, the day appointed for the dive. There were swarms of people practically pushing each other into the river trying to get on board. We couldn't figure out from where the leak had come, since we had decided there would be no advance publicity—just in case I couldn't make it.

Captain Whitman, able master of the *Antietam*, trimmed them down in short order, picking out representative reporters to cover the story, who would share it with the others on return. He chose representative still photographers and the skeletons of four newsreel camera crews. Every one of them had a wild gleam in his eye—someone was going to get killed today, and what a story that would be!

The broadcast was to be at 1:00 p.m. In time, Captain Whitman called out to his crew to shove off. The *Antietam*'s bow slipped out into the stream of the ice-choked Milwaukee River, and in a moment, we were off. Bridge after bridge swung open for us, and soon we were cutting the ripple of the outer harbor. In a few minutes, we felt the deck take life beneath our feet as our bows emerged through the breakwater into open lake.

I had planned to hit 360 feet today, the next logical increase after my previous 300-foot dive. This would be 54 feet deeper than Frank Crilley's record made twenty-two years ago and would be ample margin to officially prove the superiority of helium over air. Thus, I asked Whitman to try and anchor in as close to sixty fathoms of water as possible.

I could hear the soundings being sung out. "Fifty-seven, sir." "Fifty-seven, sir." "Fifty-seven, sir." "Fifty-eight, sir." "Fifty-eight, sir!"

We had been proceeding so cautiously that suddenly we realized there was very little time left. It was important that we be in location before one o'clock and that I be in the diving suit ready to go overboard as we joined NBC. Captain Whitman well realized this and knew that he would have to speed up his sounding operations. His vessel had been coming almost to a standstill to run each sounding. In a quick extrapolation on the basis of the apparent gradual slope beneath us, he ordered a spurt of speed to take us in one stretch to the desired depth. A sounding was taken. "By the mark, seventy, sir!" came the report. The contour had changed. The bottom had dropped off more rapidly. We had 420 feet of water beneath us.

Whitman looked at his watch then looked at me. "We have your 360 feet of water beneath us," he said, "and also 60 feet to spare." He looked at his watch again. "There is no need to go all the way to the bottom. You can stop at your desired depth. It's going to take a little time to get our anchors laid. We don't have time to look anymore."

I reluctantly agreed. He was right—360 feet of water are 360 feet of water, regardless of whether I would be standing on the bottom or just hanging on a rope. Somehow, it didn't seem quite the same—but it was too late now to search for the exact depth. If only there hadn't been so much confusion when we were ready to sail as to which reporters could go and which would have to take the story secondhand.

Over went the anchors. The *Antietam* maneuvered to feel their bite on the bottom, a check sounding was made, and the previous depth confirmed: "By the mark, seventy, sir."

Everyone was busy. The Coast Guard boys were getting out the anchors,

Ivan Vestrem (right)
dressing Max Nohl
(left) for his dive
from the *Antietam*.
MILWAUKEE JOURNAL
PHOTO

the radio boys were checking their cue channels, our diving gang was making last-minute adjustments on the gear, the cameramen were getting
ready for the kill, and I was squirming into the rubber dress.

The temperature was not much above zero. A biting breeze was blowing across the winter lake, relentlessly penetrating every stitch of clothing
anyone could wear. The lashing spray from the angry waves was freezing
on everything. There had been many massive cakes of ice through which
we had plowed coming out. In short, it was a miserable day on deck.

Except for those few who preferred the misery of the wintery deck to
the certain symptoms of mal de mer that came from stepping below, we
had all found as much excuse as possible to stay in the warm interior of
the ship. Now, however, it was time to venture out for all.

Shaking, they all, one by one, looked at me with a skeptical eye as to my
mental condition—wanting to plunge down into that iced winter water on
a day like this. In turn, at various intervals, every man on the ship finally
ventured to ask individually just exactly what I had in mind, not saying in

so many words that I was crazy but politely wondering why this must be done on a bitter December day.

The joke was really on them. While they each would suffer for me in a few minutes as I started down into the frigid waters, I knew that I would actually be the warmest one of the lot. There would be no wind down there in the first place, and I would have almost jumped overboard without the diving suit to get out of that merciless blast.

Secondly, the temperature would be considerably higher than it was up here on deck. Although to the popular mind the coldest thing on this earth is ice water, the sensation of cold is a result of its capacity to quickly conduct heat away from the body rather than its low temperature. It was probably about zero up here on deck, but that water I knew very well would not be below thirty-two degrees (Fahrenheit)—if it were colder than that, I wouldn't be going down into it except with an ice pick.

Under the diving dress, I was wearing a warm lambswool garment made very much like a baby's sleeper suit, plus several pairs of "long handled underwear," and wool socks. Unless I should spring a leak, it would be quite comfortable.

It had been decided that we would open the broadcast on deck, just as the helmet was being clamped on over my head, and that I would be on my way to the bottom within the first few minutes of our allotted time. We would have thirty minutes, and they seemed very anxious to make sure that I would be in 360 feet of water during that time. They wouldn't say so, but from the gleam in their eyes, I knew what they were thinking—they wanted to broadcast my death rattle and to make sure that they were on the air when it came.

Everyone took his station as the starting moment approached. Norman Barry, NBC's suave and handsome announcer, was to be the commentator. Ivan Vestrem, my faithful diving associate, was to take charge of tending my lines and getting me in and out of the water with the full facilities of the Coast Guard crew at his disposal. Captain Whitman was in charge of the vessel, Ken Fry was in charge of the broadcast as a whole, and Dr. End was to stand by for emergency consultation and co-commentating with Norman Barry.

Then came the fifteen-second hush signal—and except for the breaking seas, there was not a sound. Ken Fry's raised hand then dropped, and we were "on the air."

Dr. End (left) and Norman Barry (right) on the deck of the *Antietam*. COURTESY OF KATHY END

"Good afternoon, ladies and gentlemen," started the smooth genial voice of Barry, "we are bringing to you this afternoon a unique broadcast." He went on to explain where we were, the sensation of being at sea on a day like this, and what we were planning.

Frequently consulting Dr. End or asking him to explain or elaborate on some point, Barry described the experimental work we had done with helium at Marquette University, the guinea pig test dives and how they had shown us helium's effectiveness, and how we were now planning to test this suit at a depth which no human being had ever before reached.

The *Lusitania* story was told—how she had sunk in 1915 and been found during the past summer at a depth of 312 feet off the Irish coast. It was also told how, curiously enough, in that same year, 1915, almost a quarter of a century ago, the United States Navy diver, Frank Crilley, had reached a depth of 306 feet to establish the world's record for deep-sea diving, which I planned now to try to exceed.

"Are you ready, Max?" asked Barry as he finished with his preliminary explanation.

I was ready, and as I said it, I felt the clank of the tools in the last tightening of the diving dress to the helmet. Vestrem called the "Haul away!"

signal, and in a moment, I felt myself being lifted off the deck into space. I slumped down into the suit and relaxed to listen to my own broadcast. I could see nothing, since the window was badly fogged, my breath and body humidity freezing on the zero-temperature glass. Barry now started asking me questions, and I could now devote my full attention to carrying on a conversation.

Helium is tasteless, odorless, and colorless. In answer to the frequent question, "How do you know it's there?" is a suggestion for a simple and startling test. Take a deep breath of the gas in question, and then start talking. If it's helium, you will know it right away. Because of its low density, it has a decidedly different period of vibration than the same quantity of air in the larynx, just as a light, thin piano or guitar string has a different note than a heavier one. With every attempt to talk in a natural voice, out comes a high falsetto that gives a very funny effect.

Thus, the conversation with Barry was a strange one, with my helium-pitched voice and the heavy reverberations of the helmet interior.

Suddenly, I heard a gasp up on deck and in a split second felt my body splashing down into the water. The clutch on the hoist had slipped, and I had fallen the last ten feet for an unexpectedly quick entrance into the lake.

There was much excitement up on deck, which I could well hear over the telephones, everyone apparently thinking I must be dead from the fall in the huge 288-pound suit. Actually, I had hardly noticed that I was dropping, the weight of the suit giving me the inertia to plunge into the gradually retarding water with a softness that no feather mattress could boast. I bobbed up and down in the water a few times before coming to a rest with my window just under the surface.

I next rubbed the fog from the window with my "rubber nose" and saw directly in front of me the massive hull of the Coast Guard cutter.

Barry, still thinking that I was badly injured from the accidental fall into the water, seemed much concerned. This, however, is the way we usually go overboard anyway—we just jump in—and although it had been a surprise, it was not detrimental to the dive in any way.

I called, "Slack away," and heard him repeat the order to the boys on deck. I started down, the descending line running through my mittened hands.

Although this was a self-contained suit, on this experimental dive, I

Preparing for the 420-foot dive, December 1, 1937. Nohl relaxes and sinks down into the suit while waiting to be lowered into icy water, after which the buoyancy gently shoves his body upward so that his eyes are centered in the window. Helium tanks can be seen on his back.

decided to have a lifeline snapped to the helmet top. Since there was a microphone, it would be necessary to have the wire at all times anyway, and so Vestrem had previously bound the two together. I had a typical December head cold and was afraid that I was going to have trouble clearing the mucus from my Eustachian tubes. Thus I was also using a descending line, which I could use as a trolley to hang to in the event that my ears didn't clear. It would also give me something on which to hang to offset the deep lake currents.

Vestrem had apparently entrusted the actual tending of my lines to one of the Coast Guard boys while he attended to numerous other details on deck. One of the difficulties a diver is likely to have with inexperienced tenders is that they take too good of care of him, gingerly and painstakingly feeding out his lines, inch by inch, to greatly impede his progress.

The diver, actually, is always anxious to reach the bottom as fast as possible—if any retardation is necessary to save his ears, he can perform this himself either by clinging to the descending line or by lightening his suit.

"Slack away!" I called. It seemed they felt that they were feeding me down into what must look like death itself on this icy day. I went down a few feet and then found myself barely sinking.

"Slack away!" I called again, and again, and again! Finally, I paused to explain in full detail what I wanted, which was no restraint on the lines whatsoever—I wanted to control my descent myself.

This, apparently, worked. I started down like a plummet.

A pain struck through me like a streak of lightning—a "clinker" must have jammed up in my left ear. I frantically clutched for the trolley to stop myself, but my inertia was so great that, before I could come to rest, I almost lost an eardrum. The pain was so intense that I climbed back up a few feet where I immediately found relief with the lessened pressure. I swallowed, coughed, yawned, and tried everything in the book, but to no avail. Up again and down again, up again and down again I bobbed, but with no results. In shorter cycles, I rapidly plunged my body like a piston at that barricading depth, but I could not pass it however hard I hammered. I went up a few feet higher for a battering ram blow, but at exactly that same lower level, the streak of lightning hit me again and effectively made it impossible for me to go deeper.

It felt as if someone had driven an axe-head through my skull. What a ridiculous thing! On this, of all dives, my ears wouldn't cooperate.

I decided to just wait. Barry was asking a lot of questions. I selected a depth where the pain was just as bad as I could stand it and then moved an inch deeper. In this way, there would be a maximum push on that clinker. We talked about ears, helium, the *Lusitania*, and why I was not going down any farther at the moment.

I suddenly found myself viewing, through the expansive window of the helmet, a strange sight. Passing rapidly downward in the trackless blue-green water was a bright yellow manila rope. I studied it for a moment to attempt to determine what it possibly could be—it was my lifeline!

It was obvious after a moment's reflection what was happening. I had stopped my progress by hanging on to the descending trolley. I was at a depth of approximately 220 feet. When I had stopped my descent, they had continued to feel the pull on the lifeline and had kept paying it out to feed it as I had asked. They were not feeding line to *me*—they were simply feeling the weight of the 220-foot loop of manila.

I tried to explain this to Barry—that the boys were now giving me too much slack and that I wished that they would now take in the excess. In a moment, I saw the line going the other way.

We continued to talk, and I was beginning to realize that the attempt to break the record would be a failure, not because of any reflection on the helium or the new suit, but just because of a stubborn clinker in my ear. At 220 feet, I was only 86 feet from Crilley's record, but progress was now limited to a maximum of a few more inches.

Suddenly, there was a sharp click in my head. The almost unbearable pain vanished as if a fairy wand had been waved over me. The transformation from torture to what was now a crystal-clear head was as instantaneous as the wink of an eye.

I started down into the stygian gloom below. On my extended sojourn at 220 feet, I had enjoyed the last trace of deep-purple light in the water, but in a few seconds even this vanished into what might as well have been a mammoth pool of India ink.

At 240 feet, my clinker came back momentarily, and I stopped again. This time, however, it was just temporarily stuck, and a moment's yawning pushed it on its way. As I hung to blow those tubes, I felt again the drooping coils of line sagging down on me. The boys were still feeding out too much slack.

We only had a half hour on the air, and I was determined to reach 360 feet before it was over. I decided to request that Vestrem, who had had his hands overfull with other details of the diver, hereafter personally handle my lines. I must be in quite a snarl already. Vestrem agreed to take over.

I started down again, blowing every possible cubic inch of extra gas out of the dress to give myself as much weight as possible and asking for full slack on the lines.

For a moment, I thought that I was descending—purely an imaginary sensation since I had nothing with which to compare my relative movement. Grasping the descending line, which I now held loosely in the crook of my arm, I noted that it was not moving—I was standing still. There was no way in which I could increase my weight any further. I asked Barry to check whether I was all slack.

"All slack," I could hear Vestrem sing in the distance. "All slack," repeated Barry into the microphone.

Perhaps I could pull myself down as I had often done before in strong currents.

I reached down and grabbed the line as low as I could and started pulling. I moved only a few inches. Another pull, and I progressed a little further. It was like pulling a too-tight ring off a finger; though I seemed to be moving, I was slipping back eleven inches for each foot of progress I was making. I pulled harder and harder, almost frantically lunging at the line. I was just about in the same place five minutes later. I was also panting.

Perhaps I could climb up a little ways and find the snarl that was holding me. I reversed my efforts and found that I could not move up. I tried to move down again and then up again.

I was stuck—stuck at 240 feet in total darkness. I was exhausted.

During these struggles, I had tried to carry on a conversation with Barry but had been so busy that I had not had much to say other than to answer his questions. He had been warning me that we were running out of time. And now I heard the words that I had feared, "We regret, ladies and gentlemen, that our time is drawing to a close and that we will have to conclude this broadcast with Max Gene Nohl at a depth of 240 feet on his attempt to break the world's record for deep-sea diving. We are broadcasting from the United States Coast Guard cutter *Antietam*. This is the National Broadcasting Company." And then over the phone came, "We're off the air—what's wrong?"

29

Six Hundred Thousand Pounds

I had a pretty good idea of how everyone up on deck felt. I hadn't even closely approached the record, no one had been killed, and all in all it had been a dull day.

There was nothing that I could do about it. I couldn't move. I tried to explain my plight and then nestled back into the diving suit to relax and sulk. Those were about the only two things that a person could do in this three-dimensional inky nothing.

They tried to pull me up, but all hands combined couldn't budge me an inch.

Captain Whitman then took the situation in hand. To the lower end of the descending line, he had fastened a huge iron weight. This would be almost impossible to lift by hand, but it could be done on the power winch. The line was led to the winch bollard, and soon everything started moving upward.

I saw the dim glow of light returning to the water, assuring me that I was on the way back. Whitman had the boys untangle the two lines as they came out of the water. I saw some of the pictures later that had been snapped of the snarl, and it was then no wonder I had been trapped.

It seemed as if an eternity had passed before I saw, looming up to my right side, the massive under hull of the *Antietam*. A moment later, a glare of light struck me as the window was uncovered in the trough of a wave. Within a few minutes, I was standing on deck.

I cleared the rapidly fogging window and peered out. The deck was covered with snarls of rope, which the Coast Guard boys were rapidly straightening out. Standing around were a score of long, long faces dis-gustedly watching the ice forming on my dripping suit.

Until this moment, I had not understood exactly what had happened. They explained. When I stopped to clear my ears, the boys had kept on feeding line into the water. The loose coils of slack had fouled me and my descending line—they had pulled them taut and slacked them off several times to make a series of snarls that looked like a sailor's nightmare.

I had been about an hour in that ice water and was beginning to feel it—my body was cold and yet was still dry. I asked to talk to Vestrem on the mike.

"Is there any reason why I can't try it again?" I asked.

Vestrem put it up to the gang. End said, "No." Whitman said, "No." There was no reason, however, other than we had failed the first time, and I looked as though I should be very cold. End came to the microphone and talked about decompression. "How can we calculate a decompression for another dive," he asked, "when it will be a cumulative effect of a 220-foot dive, a 240-foot dive, and whatever you might do the next time?"

He was right. However, I knew as well as Dr. End that we were really so far out on the limb with our theories that another factor of possible error wouldn't make much difference. We had assumed theoretical character-istics for the gas helium and were very accurately calculating our decom-pression time on the basis of these. By going down again now, there was still a certain amount of gas retention from the first dive to be included—but he could pretty well guess at that so as not to deviate too far from our purely assumed theory.

This would not, then, be under perfect theoretical conditions. How-ever, that day might never come. Could we ever get this magnificent setup again—a Coast Guard cutter, Dr. End, a large crew of men, and the publicity opportunity that would be of such help to us in our future work?

I called Vestrem to the microphone. Although we were standing only a few feet apart, I was still locked in the 288-pound suit and communi-cation was almost impossible except by telephone. It seemed strangely silly to have to phone someone whom I could touch by extending my hand.

"You understand that I, as diver, have the final decision," I explained to him, referring to a basic code of our profession with which he was well familiar.

"Yes," he admitted.

"I want to go overboard again—right away!" I requested.

Captain Whitman knew that this was my authority or that of any diver in gear. In a moment, I felt myself being lifted up off the deck and swung out over the rail. A few seconds later, my feet hit the water, the icy surface crept up over my body, and then a wave crest swept over my helmet, and I disappeared for the second portion of my dive.

Vestrem and I now had our system working, and he was going to personally feed my lines.

I started down like a plummet, Barry calling out the time and depths to me.

"Forty feet—thirty seconds," came his suave voice. We were not on the air, and there would be no conversation other than that strictly necessary for the execution of the dive.

"One hundred feet—one minute and ten seconds."

"One hundred and fifty feet—two minutes and five seconds."

"Two hundred feet—three minutes and ten seconds." The water was getting almost blue black, the violet end of the spectrum that was the last to be filtered out. The last trace of light was now discernable, being so faint that I was not quite certain whether it was there or not. Soon, I was in total darkness.

"Three hundred feet—five minutes and fifty-five seconds." My rate of descent was diminishing somewhat, probably because of the pull of the tremendous expanse of lifeline. However, I was going as fast as I could sink, my ears working like a charm.

"Three hundred and six feet," sang out Barry, "Frank Crilley's record!" A funny little feeling shot through me.

"Three hundred and twelve feet," added Barry before he had barely finished the previous call. "The *Lusitania*!"

"Three hundred and twenty feet." I paused, mentally, for a moment. I had accomplished my mission by a fair margin and could now quit if I so chose. However, my head was clear as crystal, my breathing was relaxed and normal, I was not conscious that I had a heart, and I felt like a million dollars—or perhaps like twenty million dollars, the reported treasure in the *Lusitania*. There was no reason why I should not keep going on to attempt to reach the proposed goal of 360 feet.

"Three hundred and forty feet," sang out Barry, "seven minutes and five seconds." Everything was still fine.

"Three hundred and sixty feet—HOLD HIM—HOLD HIM—HOLD HIM," screamed Barry, and turning back the microphone, he added, "seven minutes and thirty seconds—you made it!"

I felt my progress arrested. I was at my planned depth.

Everything was working like a charm. I again checked my breathing, heart, and state of consciousness. I felt as fine as I ever could.

Something was swelling up inside of me. Here I was in 360 feet of water—deeper than any human being had ever gone in a diving suit. But under my lead shoes was only another sixty feet of water to the bottom.

I had long known and experienced what the restricting factors were to deep water—that struggle just to stay conscious and that vague feeling of incoordination. There was no trace of any of this. The thought of trying for that last 60 feet had never seriously entered my mind since Captain Whitman advised me that we had 420 feet beneath us. I was probably too much worried about the first 360 feet to even consider what was below that.

However, that first 360 feet were just a memory now. Inside of me was growing the full fury of that relentless urge that had driven me through all of these years down into the mysterious depths of the sea. I was, frankly, afraid. The water was so black that I found myself mechanically opening and closing my eyelids to attempt to determine whether or not they were open. The weird roar of my breathing reverberated like that of a fantastic mad monster in a sealed sepulcher. I was shaking with cold. And yet, fear transformed into fascination was urging me downward. It was the same compulsion that I had always felt as my body slid down into dark deep water. But now, at this heretofore unpenetrated depth, it was roaring in my brain, "Go on!"

"Slack away!" I called over the phones.

"But, Max, you—" protested Barry.

"Slack away," I repeated, "you know the law."

The restraining line loosened, and I started down.

"Three hundred and eight—four hundred," called out Barry, "don't you think—"

It was too late. My lead shoes sank into a soft oozy mud.

"Four hundred and twenty feet—nine minutes and five seconds," called Barry.

"I'm on the bottom," I answered.

I started crawling around in the ooze on my hands and knees. It was so black that it was difficult to tell where the water left off and the gooey bottom started.

"Listen!" called Barry, as he held the microphone as high as he could reach. The ship's whistles and horns were all blowing. A warm glow ran through me as I vaguely heard the din on deck.

I heard them say that I was under a pressure of 192 pounds per square inch. That was about what a locomotive might carry in her boiler under full steam. If I had a pipe to which I could press my lips, the upper end being connected to the high-pressure piston of a locomotive, I could blow into it and start her down the track with a train of cars like a scared jackrabbit.

My body, as I recalled, totaled over three thousand square inches. At 192 pounds per square inch, I was supporting a total water weight of six hundred thousand pounds—the weight of two hundred automobiles.

As startling as these figures seemed, however, it was quite comfortable. The pressure was hydraulic. It came from every direction. There seemed to be no undesirable effect on the various fluid tissues of the body.

There were quite a number of observations that I was anxious to make, and I explained to Barry to please not call me unless necessary. I wanted to run through a series of mental gymnastics and concentrated introspection so that I might later have some basis on which to base my conclusions.

"Two times two is four, times two is eight, times two is sixteen, times two is thirty-two," and I went on, going up into the thousands.

My mind was clear. I called Barry and conversed with him for a minute. "Do I sound all right?" I asked him.

"Sounds OK to me," assured Norm.

I concentrated on how I was breathing for a minute. I tried to analyze every part of my body of which I could be conscious. There didn't seem to be anything different from what it would be in ten feet of water.

"Hey, Max," called Barry, "we are going on the air again."

Back on deck after the record-setting dive. Nohl's football helmet is for warmth and protection from injury in the event of an accidental "blowup" into the bottom of the steel mother vessel. Ivan Vestrem (stocking cap) and Don Chase (far right) start undressing operations.

"How come?" I asked.

"They were so worried about leaving you down there in 240 feet of water that the New York office asked permission to cancel a commercial program—we have fifteen more minutes starting in four minutes from now."

I had finished every possible observation that I could make by that time, and then I heard the genial voice of Barry: "Good afternoon, ladies and gentlemen, we return to the deep waters of Lake Michigan."

After Barry finished his explanatory remarks, we talked for a few minutes, and then I decided to start upward for a good long decompression in comparison with the theoretical time that we should figure from the accumulation of the two dives.

Dr. End had been working frantically with the slide rule all of this time, designing a new set of decompression tables for the unexpected depth which I had reached, and I asked that he give ample factor of safety to these. My first stage was at two hundred feet, to which level I went directly from the bottom. It was at this stage that the broadcast again ended and the

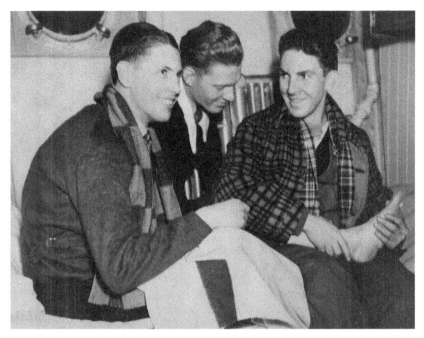

Nohl (left) relaxes with Vestrem (center) and End (right) after record-breaking dive.

dive was over except for a long dangling process in midwater. Four hours after I first entered, my helmet broke the surface, and I emerged from the water, squinting at the bright light and dripping the last scurrying drops that tried to escape before freezing.

Three-quarters of our earth's surface is ocean floor—today, an almost totally unexplored world. In this new magic gas, helium, was the key to the conquest of the last great frontier. It worked!

30

Singing for My Dinner

Johnny Craig wired his congratulations from someplace in Iowa where he was on lecture tour. The *Lusitania* was "in the bag."

However, this was early December, and diving off the Irish coast would not conceivably be before May or June. Since the completion of the first model of the Super Suit last spring, I had not used any of Craig's capital with the thought that if the Lucy should fall through, it would weigh too heavily on him and he alone would be the loser. The new developments on the suit and the long, expensive preparations for this dive had been coming out of my dwindling reserves, and I was very rapidly approaching being broke.

If, back in my senior year in high school, someone had asked me to get up and say a few words before a crowd of people, I would have much preferred to take on, single-handed, a blood-crazed school of tiger sharks. At the time of the Seth Parker Expedition four years ago, following a flood of publicity, I finally had consented to give a little luncheon talk before the Milwaukee Optimist Club—my father had been president of both the local and international organization. I had been scared stiff but found that, once I got started, there was nothing much to it.

Two people in the audience had also belonged to other service clubs, and when I returned home that night, there were calls waiting for me. Would I repeat the same talk to their organizations? Each of these, in turn, had led to several others, until the progression had grown to such proportions that I could no longer handle it.

In the past years, I had gone quite deeply into motion picture photography and now found that I had quite an extensive collection of films of our various expeditions. I had talked to Johnny Craig, now an internationally known and well-liked lecturer, and asked him what I should do.

"Soak 'em," said Johnny, giving me sketches of some of the fancy fees he was collecting.

"I can't do it—I just can't stare a man in the eye and tell him that it will cost him fifty or a hundred or more dollars just to listen to me talk for an hour," I said, revealing the dilemma I had been in for all of this time.

"I'll tell you what," gleamed Johnny, "it's something we'd have to do anyway."

I was listening.

"We have our little company. We are going over to the Lucy. We are going to do lots of other things. There will be a lot of publicity." He continued, "I am planning to lecture for a long time. If you go out on the same expeditions and come back with the same film and run around showing it to anyone who will look at it, how do you think my lecture bureau will feel? What do you think will happen to my lecture business?"

He was right. We drew up a scale of minimum prices for which I could appear. This would be fine if only I had the courage to quote them.

The phone rang that evening. It was long distance. "How would you like to tell us some of your experiences?"

I was very flattered and all that sort of thing, but I was now, I explained, "under a new contract and will not be allowed to make any public appearances except as provided and on a specified fee basis." I gulped, but it was out.

The voice insisted on going on. "How much would that be?"

"For a luncheon engagement with pictures, fifty dollars and expenses." That would end that.

"How about next Tuesday? The fee is perfectly all right!"

I couldn't believe it, but next Tuesday came and went, and I had the check in my pocket.

The fee didn't seem to make any difference to any of them except an occasional little church group or small club with no budget for professional speakers. Business was good.

It was getting so good, in fact, that I found I was gradually getting away from the diving and devoting my time to the terrific amount of work in booking, routing, and taking care of the myriad of business details associated with the profession.

And then one day, I was asked, "How would you like a contract with the Redpath Bureau for a transcontinental lecture tour?" They would book the entire tour and handle every detail for a percentage of the gross.

James Redpath had organized this bureau bearing his name in 1868, and their list had comprised outstanding national celebrities ever since. They had been particularly famous for their "Chautauqua," a traveling tent show that had brought a homespun mixture of drama, religion, comedy, music, and entertainment to every town and hamlet in the country. As a young boy, I had worked for a few days with a Chautauqua, and its memories were vivid in my mind as I walked into the Redpath office, now exclusively a lecture bureau since the tent shows had been forced to give way to the advent of the radio and the motion picture.

They had their main office in Chicago with reciprocal offices in Pittsburgh and New York and exchange contracts with the Dixie Bureau in Dallas, Texas; the Ellison-White Bureau in Portland, Oregon; and the Alkahest Bureau in Atlanta, Georgia. This sounded like a good coverage of the country, and I was eager to get started.

Before they risked my talents too far on their clients, however, in spite of testimonials that I had from my own bookings, they wanted to hear me and see with their own eyes whether I could "deliver the goods." There was to be a banquet of the international Lyceum Association in Chicago, and I would be booked as the feature attraction of the evening. The audience would consist of members of the association—speakers, entertainers, members, agents, and salespeople of many lecture bureaus, and chairmen of many clubs who booked attractions. Carl Backman, manager of the Chicago office, explained that this would be an excellent chance to present my show for consideration directly to these people.

There were a number of other acts on the program that night, each of these not being allowed more than five to ten minutes. The fellow on before me was a magician who could produce almost anything out of a hat. I caught a glimpse of the first few minutes of his act and was utterly mystified to see him bring (one by one) a flock of ducks out of the strangest places. I watched a few minutes more then decided it was time to go into the backstage dressing room and give my hair a last-minute combing.

The light was off in the small cubicle, but I remembered from a previous visit that it was operated by a dangling string in the middle of the

The front cover of Nohl's Redpath publicity brochure.

room. The spring door slammed behind me, and I found myself in as total darkness as had existed in the deep waters of Lake Michigan.

Suddenly, there was a din a few feet in front of me as though all hell had broken loose. A turbulent spray of water was filling the room. Had I stepped into the shower and turned it on, with a thousand wild animals screaming at me in their amusement? I groped frantically for the light, realizing that I was being soaked in my freshly pressed lecture suit. I couldn't find it. I tried for the door but was now completely lost in the pitch-black void. The light would be the quickest. I tried again and, in a moment, had my hand on the dangling cord.

I pulled and there before my eyes was my competitor's unintentional sabotage plot.

The washbowls were each filled to capacity with water, and there, screaming and flapping in the water, was the flock of ducks.

I was soaking wet and half scared to death. I opened the door. Booming over the microphone was coming "—and now it is our pleasure to present a man who has just recently broken the world's record for deep-sea diving, a man who—"

I had to run to make the stage as my introducer turned to the wings and said, "and now we present Mr. Nohl."

I was cool and refreshed and had forgotten all about my nervousness for this critical tryout lecture. Other divers had probably appeared before audiences and told the story of their experiences. This probably was the first time that a diver had appeared on a platform in a tuxedo still dripping wet from his last submersion.

My forty-five minutes went off smoothly, and I was now in the lecture business.

31

DESCO

If I had heard the question once I had heard it a thousand times. I had a stock answer, and I had given that answer a thousand times, too.

The question was, in essence: "Why don't you manufacture the new diving suit?"

The answer was, in essence: "We developed the new equipment for our own use on our own expeditions—the rewards are lying down at the bottom of the seas, not in the necessarily small-scale manufacturing and merchandising of the product."

It was true that we had "a million dollars' worth of free publicity." A vast amount of copy had been printed in almost every newspaper and magazine in almost every language throughout the entire world. The equipment had proved its worth. This certainly was a fine foundation on which to base a business.

The actual statistics were rather grim, however. I was well familiar with them and based my stock answer on these facts.

In the entire United States, I estimated that there were only about one hundred professional divers—men who made their living exclusively from the diving business. This might have even been a high estimate. I thought that I knew, or knew of, most of them. There were others, of course, who were part-time divers, thrill seekers, hobbyists, students, and a few more, but the list was amazingly small.

To supply these few men, there were essentially only two firms in the United States that manufactured diving equipment. One of them, the Andrew J. Morse Company in Boston, which had been so kind to me years ago as I got started in the business, had been founded in 1837—just exactly one hundred years before the date of the incidents I am now describing. The other company, A. Schrader's Son in Brooklyn, New York, had been founded seven years later in 1844.

In both cases, diving equipment was not the company's only business. Morse also manufactured a fine line of fire engine equipment, the beautiful shiny brass nozzles and fittings on most fire engines throughout the country being its proud product. The Schrader diving department was merely a very tiny subdivision of the large Schrader Valve company, which makes the well-known little valves used on most automobile tires. I had visited the Schrader plant numerous times and was surprised to discover that many of the employees didn't even know that they made diving equipment. However, all divers did, since the relative merits of Morse and Schrader equipment was a hotly contested subject of conversation at many a diver's bull session.

Since the original invention of the diving suit in 1837, at the time that Morse was founded, there had been practically no change in the item's design. Very unlike fashions in women's clothes and automobiles, the 1937 model of the suit was almost indistinguishable from the 1837 model. It is true that the gasoline engine–driven air compressor had largely replaced the man-driven hand pump as a source of air supply and that optional telephones had been made available to these divers who wanted them, but other than that, diving equipment had made little progress.

The worst part of it was that most of the equipment just wouldn't wear out. The helmets were made of copper and brass and were totally unaffected even by a passing century. The lead belts and lead shoes were not harmed by the passing years. The rubber portions of the equipment did deteriorate, but replacing them was relatively inexpensive.

Thus, as a prospective manufacturer of diving equipment, the prospects looked very dismal indeed. In brief, we had a potential one hundred customers. They all had all of the equipment that they needed, and they would sell it to their successors when they retired. The two companies that did make diving equipment did it in conjunction with other businesses, and these two fine companies each had almost a century of tradition behind them, with which it would be mighty hard to compete. It was no wonder I had a stock answer, "No!" for the many suggestions.

However, Jack Browne didn't quite agree with me. He viewed the matter from another angle. At first, I didn't quite understand his firm convictions that we should go into manufacturing. For one, Jack wasn't the manufacturing type. He knew nothing of industry and business. He had had no related college training. Jack was a diver—and not just another diver, but one of the finest, most naturally qualified divers in the world, I believed.

Jack had that peculiar combination of physical stamina, unlimited courage, and mechanical genius that it takes to make a fine diver. But above everything else, Jack had a cold, calculated spirit of adventure that I had found unrivaled in any other person I had ever met. I just couldn't picture Jack directing his energies to the cost accounting of a production line problem. But Jack kept talking about manufacturing.

I had been very stupid about it all. One day, the light came.

Jack was head over heels in love. He was so much in love that it was pitiful. He couldn't eat. He couldn't sleep. He was bumping into things. His work, as we continued to build the equipment we would need for the *Lusitania* operations, was slipping badly.

Yet, Barbara didn't want Jack to go halfway round the world to salvage the *Lusitania* or to salvage anything else. Barbara didn't want a million dollars—she just wanted Jack and a home in Milwaukee and a normal married life.

Jack, either consciously or unconsciously, felt that manufacturing was the answer. He could still stay in the diving business, and yet he could get married and could establish and maintain a home. The difficulties above, which I outlined to him, were just details. We would lick those problems as they arose.

The self-contained suit with which I had just made the 420-foot dive was fine for its intended use in salvaging the *Lusitania*. It was really a heavy-duty piece of apparatus. It had every possible safety factor and luxury that could be built into a diving suit. We would be diving from the *Orphir*, a big ship with everything in booms, steam winches, and husky equipment, and a few pounds more or less in the suit wouldn't make any difference. We would be lowered into the water by boom and lifted from the water by boom and steam winch. Underwater, the extra pounds would just provide the extra traction that a diver would want in the tide currents of the open Atlantic.

The old and the new diving suits, with purely "accidental" gestures.

However, the big suit would not be ideal for the average commercial diver operating from a small boat or dock. He would not need all of these pounds nor all of the luxuries that we had incorporated. What we wanted, if we were going into commercial production, was a decidedly simplified, low-cost, lightweight outfit. Jack and I thoroughly agreed on all of the many details that we would incorporate in the new commercial model—if we should form a manufacturing company.

One day, Jack didn't show up for work. That was nothing new, since anything concerning Barbara took precedence over the job of preparing for the forthcoming expedition. The next day, he telephoned and told me that he was afraid he was going to have to quit. His father was sick.

His father had been sick for some time. I asked Jack if he had taken a turn for the worse, but Jack didn't think so. He was just "still sick."

The next day, there was an article in the *Milwaukee Journal*. It was very short and was written by someone who had quite a sense of humor. The article read:

JACK BROWNE QUITS "LUSITANIA EXPEDITION"

Milwaukee, 1937. Jack Browne today announced he had resigned from the "Lusitania" expedition. He stated that he was leaving because of the ill-health of his father.

Mrs. Norman Kopmeier today announced the engagement of her daughter Barbara to Jack Browne.

I didn't see Jack for several months. One day, I was stopping in at one of our suppliers of materials used in the diving suit, and I was asked, "What's all this I hear about Jack Browne starting up a new company for the manufacturing of diving equipment?"

"Oh, really?" I asked, but then I knew. I called Jack right away.

"What's all this I hear about you starting up a new company?" I bounced right back at him.

"Well," he said, "I knew you weren't interested in manufacturing. I gotta do something—and you know why."

"What do you say we talk it over?" I suggested, and Jack very graciously agreed.

Jack had, in the meantime, gone to Norman L. Kuehn, owner of the Kuehn Rubber Company of Milwaukee, and sounded him out. Kuehn had been intensely interested in our diving. I had originally gone to him for the construction of the diving dress that I had just used to make the 420-foot record. Kuehn had shown much interest all of the way along, far beyond that which pertained to the mere rubber problems in the development and sale of that suit. He had offered consultation on many questions as they came up. He had done much experimental work for us without billing us for it. He had made countless inquiries around the country in an effort to get the right rubber process for us and had finally come through with just what we wanted. On the day of the 420-foot dive, he was one of the chosen few who were given press passes to go out on the Coast Guard cutter. He had taken his 16 mm camera along and had shot hundreds of feet of beautiful film and had given me a copy of it all. I felt that Kuehn really was interested.

Everyone was somewhat embarrassed. Jack had rightfully assumed that I was not interested in manufacture. But—I had a very comprehensive patent on the suit, in which piece of paper alone I had invested almost two thousand dollars. If any attempts were going to be made to commercialize it, I certainly wanted to be part of it.

It was unanimously agreed that we would join forces. Kuehn would be the "angel" and would supply the finances and facilities of his rubber organization; Jack would be the production department and see that the equipment was built; and I would be the sales and promotion department. This was a fine setup for me, since I could advantageously combine these functions with my widely publicized expeditions and also with my lecture and motion picture work with mutual benefit to all. It certainly was worth a try.

I was much inspired by the enthusiasm of a successful and practical businessman like Norman Kuehn. He believed that we could "crack the market."

The next two years were trying ones for our new little company. It was very difficult for all three of us.

Jack had the biggest problem of all. DESCO, our Diving Equipment & Supply Company, had to have an income before Jack's salary could be paid. He was working on the understanding that no income for DESCO meant no salary for Jack. We all knew, with the possible exception of Jack, that it would be a long time before we would have money coming in, even though Kuehn had agreed to let Jack draw up to the total of gross receipts.

My problem was no problem at all. I merely had to keep right on doing what I had been doing but with the thought that I would slant things toward DESCO as much as possible. On my lecture tours, I would be showing audiences throughout the country the new equipment and, without undue imposition on the sponsors, work in a careful plug for DESCO. On my expeditions, I would publicize and use DESCO equipment. I would get no salary from the company, and I didn't want one—I would get a contractual commission on the gross sales, which would reward me as I built up the business.

Display of the "Nohl-Browne Diving Gear" at Kuehn Rubber Company.

I probably set another new world's record in the next couple of years—a record that would stand forever, a record that would actually be impossible to beat. I established a new record for low volume of sales.

DESCO sold absolutely nothing. DESCO's gross sales were zero dollars. No one could possibly do worse.

The circumstances were actually against us. I had a lot of potential customers on the string who were very much interested in the new suit. However, they were still skeptical about buying something as new and as unproven as this from a company that was obviously new and had no evidence of stability. The suits were essentially handmade and were thus quite expensive. The potential customers would thus have to make very substantial investment, and it is not difficult to understand their reticence.

The suits were so expensive that Kuehn very understandably didn't want to put a lot of money into them until he was quite certain that there was a market. He had poured several thousand dollars into just the first two models, which Jack and I had used for testing. There had been so many alterations and design changes that these were now very sad-looking pieces of merchandise. Kuehn felt that we could use these as demonstrators and then build brand spanking new ones for deliveries. However, the demonstrators were by now such tell-tale patchworks of scores of ideas and counterideas that they looked like something one would expect to find on the scrap pile. I was convinced that any prospective client who just looked at one of these experimental monstrosities would quickly decide that he didn't want to do business with any such company.

It was, in short, a stalemate. I couldn't sell without something to show; Kuehn didn't want to go in deeper without assurance of sales; and Jack, sans salary, was in the middle.

I went to Washington and presented our literature to the Navy Department, but no purchase orders were going to be issued until elaborate testing had been completed. Kuehn, and I certainly understood his feelings, quickly terminated this with a "No."

DESCO was in her death throes. Jack had to go out and get a job. There was no dissolution—she just was put up on the shelf. I forgot about it and went on with my diving business.

DESCO didn't die, however. She was just hibernating. A war was coming which I, along with hundreds of millions of other Americans, didn't believe would come.

I dropped out of the company in 1940. Little did I then know that the spark of life was soon to come to the sleeping DESCO—in fewer than five years, she was to grow to be the largest manufacturer of diving equipment in the world. At the time, however, I believed DESCO was as dead as a dodo.

Also, the *Lusitania* now looked hopeless. From month to month for the past year or two, there had always been hope. However, Europe was now in a state of war tension that would make North Atlantic salvage operation next to impossible.

The mighty ship, now lying in her deep grave, had been the source of much international controversy. She had sailed from New York on May 1, 1915. Although the United States was not yet involved in conflict, a state of war existed between Germany and England.

Germany was very confident that the *Lusitania* was carrying in her massive holds a cargo of munitions and supplies for the Allied troops fighting in France. If this was the case, she would automatically be declared a vessel of war, and the Germans would have the right to sink her in any way they could.

The British government, on the other hand, was equally vigorously contending that the Lusitania was carrying no such munitions or supplies for the Allied troops and that she was positively *not* a vessel of war.

The people in the German government apparently were certain enough of their contention that they had made elaborate and exacting plans to sink the ship. However, they were also aware that on board the mighty Queen of the Seas, there would be approximately three thousand men, women, and children who were perfectly innocent passengers. Injury or loss of life to these people, 159 of whom were Americans, would certainly stir up a great amount of unnecessary ill feeling and might even draw America into the war. However, the Lusitania must not deliver her cargo to the Allied troops. One of the most highly prized trump cards in the German hand was her ability to blockade the sea-lanes to the British Isles and France through the Kaiser's powerful U-boat fleet.

On April 22, 1915, not long before the sailing of the great ship, a large, conspicuous black-bordered box was placed on the front page of both the New York Herald Tribune and the New York Times, clearly warning every potential passenger that the Lusitania was, in the eyes of the German government, a vessel of war and that she would be torpedoed if the opportunity presented itself. To be positively certain that there was no misunderstanding, a representative of the German government stood at the gangplank of the Cunard liner as she was loading and personally notified each and every passenger of the controversial status of the ship and their intentions to torpedo her.

There was little attention paid to these warnings. Americans, in their traditional isolationistic attitude of that time, laughed, and as much as said and believed, "It can't happen to us."

Six days later, two waiting torpedoes crashed into her starboard side as she steered a zigzag course along the coast of Ireland. An estimated 1,198 men, women, and children went down to their watery grave with the great Queen of the Seas.

In 1938, no diver had ever been down to her hull, lying just six feet beyond the world's deep-sea diving record. But the world would be very much interested in what she really was carrying in her cargo holds.

The Lucy had been located on October 6, 1935, the day after the fishermen's party in the little pub in Ireland. I made my 420-foot dive on December 1, 1937. We were all packed to leave for England in the spring of 1938 but were delayed week by week by restraining cables from England as they waited for various political and internal difficulties to straighten themselves out. The war tension between England and Germany was at a peak. We were playing with a very touchy matter when we revealed that we were in full readiness to reach any portion of the questionable ship's hull.

We filled in each new day as the summer wore on with the construction and testing of new diving equipment and underwater cameras. However, on each of those days, we labored under the thought that we had to be ready to pack everything into a box and to leave that very night should the anticipated cable arrive. When I finally tore the July page from the calendar and August actually arrived, I knew it was now too late. The fury of the North Atlantic unleashed its wrath in early September, and it would be folly to start now. It would now definitely be the spring of 1939.

Ivan Vestrem (left) and Nohl (right) test cameras.

32

I Am Theoretically Dead

What, really, were the bends? The more we studied them, the less we seemed to understand them.

A friend of mine, a diver from the United States Navy, had one night gone to a dance. About midnight, in the middle of the floor, as he trod the light fantastic, the room suddenly started to swim. This wouldn't do, he thought, in the arms of his lovely lady. He blinked his eyes. He must stay conscious. People would think that he was drunk. He blinked again. What were those bars in front of his eyes? The round port through which he was looking? The strange clammy feeling of oppression over his entire body?

He blinked once more. Suddenly, he was fully conscious. He wasn't dancing anymore. Where was his girl? The couples? The dance floor? He was in a diving suit! He was in 150 feet of water, still wearing his tuxedo!

This was no dream. He was really down there. He had really been dancing, too. There was just a lapse of time between the two that he could not recall.

In the middle of the dance, he later learned, he had suddenly slumped in his partner's arms and dropped to the floor like a wet dishrag. "Those navy boys drink too much," was the first reaction of everyone who disgustedly saw him limply lie there. His diving pals knew better.

He hadn't had a drink that day. That afternoon, they had been running some deep-water experimental dives, attempting to learn by simple trial and error the minimum cutting to which the classical decompression tables could be pared.

He had been down in 150 feet of water for an extensive exposure and had then been hauled up to the surface on the bare skeleton of a decompression schedule—only a portion of the standard specified time—in order to learn how far a man could go in an emergency.

His friends had given him, that night, the only treatment known for that insidious disease—recompression. By putting him back under the same pressure at which he had caught the disease, the bubbles in his blood were compressed to a small portion of their size, and the capacity of the blood to redissolve the gas was greatly increased. Apparently, it worked, for here he was, fully conscious and feeling fine in the diving suit. He now had only to go up cautiously and pass those bubbles slowly out through the lungs as the pressure again fell off.

The bends had hit him at midnight, seven hours after he had emerged from the water. We knew of many other similar cases. Yet, the bottle of soda pop doesn't wait seven hours to fizz—it starts the minute the cap is removed. Perhaps bubbles were not the only explanation of the bends.

We had studied some of the reports of the incidence of the bends among tunnel workers. These men must submit their bodies to the tremendous pressures of air with which subriver tunnels must be filled.

During the drilling of the Hudson tubes, the cases of the bends had reached alarming proportions, running up into the thousands.

The so-called sandhogs were dying like flies, and it almost seemed as if the whole project would have to be called off. Increasing the decompression time didn't seem to help. Nothing seemed to help.

A visiting engineer suggested that a pressure ventilation system might possibly be of assistance, since the air was rather foul—that is, high in carbon dioxide content. In desperation, the equipment was installed.

Their problem disappeared as if by magic. What had carbon dioxide to do with the fizzing of air bubbles in the blood? Perhaps here, again, was an indication that bubbles were not the only answer.

Down in Tarpon Springs, Florida, there had always been certain diving vessels that were known as jinx boats. "The diver, he all time get killed," the Greeks would say. "Boat, she bad luck." They had superstitious reasons for these jinxes—the boats had been launched on the wrong day or the captains had committed some sin.

I had a hunch that there was some more tangible explanation than that a boat had sailed on her first cruise on a Friday, rather than on a Monday. Perhaps there was some relationship to these cases of the bends and the sandhogs in the tunnels. If the divers from the jinxed boats did not have enough air to properly ventilate their suits, they too would have suffered from excessively high carbon dioxide. If they did not have enough air, it would probably be because they were diving with too small air compressors. *If I could just see the equipment on these jinx boats*, I thought.

It was true! They were all fitted with undersized or too slowly turning compressors!

There was the mysterious ghost—a real and tangible thing. There it was—but to see it was not to understand it. Insufficient supply of air could probably cause the bends—but why? Fizzing wouldn't explain that.

Why, in addition, did the Greeks collapse with this disease when they were excessively tired, run down, or had been overworking? This was an experience I had even noted myself in my own diving, not having been killed so far, but definitely having noticed "diverse pains," the first and mildest stages of the bends, on occasions when I had dived under these conditions. Perhaps it was something more than fizzing.

There was no question that the bubbles were a contributing factor to this complex disease. There was something that was now very clearly an even greater factor. It was the question of the acid-alkaline balance of the blood—the phenomenon of agglutination of the red blood cells—and the insidious effects of the so-called inert nitrogen under pressure, which is four-fifths of atmospheric air.

It seemed so very clear to me that this was the answer and might explain all of the hitherto unexplainable aspects of this mysterious ailment.

There was one way to prove it. If I could submit my body to the pressure of deep water for many hours, and if it were possible to take blood samples periodically and subject them to laboratory analysis, it would be possible, by a study of these findings, to determine whether our theory was right. If the acidity (the pH) of the blood increased, we were correct. If it did not change, we were off on the wrong tangent.

Dr. End, in a diving suit, would have a difficult time drawing blood from any of my veins while I was in another diving suit. However, in the great tank that we had used to pioneer our first helium dives, we could conveniently do this. Dr. End could go right in with me under pressure. We could have light, warmth, food, water, blankets, cots, and even a book to read.

Under the hull of the *Seth Parker* in Norfolk Harbor, I had spent a couple of eight-hour days in the water but just at very moderate depth at the level of her keel. No diver, to our knowledge, had spent any more than eight hours in the water in one day, and in all such cases, they had done so at similarly shallow depths.

What would happen if we spent eight hours in deep water? If the "bubble theory" were correct, after approximately three hours we would become completely saturated with nitrogen and longer exposure would not dissolve any more gas into our blood and tissues. Thus, any hours after three would be "free"—the decompression for eight hours would be the same as for three, and we would have a net gain of five hours of useful work.

The advantages of this, if it worked, seemed almost fantastic. We could sink a caisson down on a wreck and could live inside of it. Like a huge open helmet, it would be open to the water on the bottom through a trapdoor. It could contain beds, goods, and everything that we might need. From here, we could venture out on the wreck in alternate periods of work and rest, returning to the underwater house without delay for sleep, tools, hot food, or whatever we needed. There would be no decompression except for the final return to the surface, which might not be for days or weeks. We would be free from surface storms. It would all depend on the answer to one question: what body changes occur under long exposure?

We would find out. I would move into the tank and live there for three full days—seventy-two hours under a pressure equivalent to one hundred feet of water (44 pounds per square inch, or a total pressure over the surface of my body of 132,000 pounds).

Dr. End was enthused as to what we might learn from the proposed experiment. Joe Fischer, papa of the chamber, was not only willing—he eagerly volunteered to get a cot and eat and sleep next to the control board for the three days.

Dr. Eugene Smith, director of the hospital, was willing—provided we would sign a release vindicating them from any responsibility in case things didn't go well. Little did he realize how glad he would be later that he had this release.

We packed sandwiches, clothes, medical instruments, books, thermos bottles, fruit, and anything that we might need for the three-day stretch.

Dr. End was going in with me but would determine the length of his stay as the results of the first blood tests were telephoned back to us.

We had planned to have no publicity on this test, particularly since we were very dubious as to its outcome. Arriving at the room in the base-ment where the giant silver cylinder was reposing, we were surprised to find the place so packed with reporters and photographers that we could hardly get in. There had been a leak. However, they finally agreed to treat it with the scientific respect that we wanted, and so we decided to go ahead anyway.

This time, we were not planning to breathe helium but just good old-fashioned air. The helium had thus far worked so miraculously that, in a sense, this experiment was a backtrack to find out what the bends really were. In this way, we could better explain the overgratifying results and more judiciously understand their future use. Thus, we had to wear no masks but just breathe and live in the tremendous pressure that Fischer was about to pour into the tank.

We closed the door and started to unpack our suitcases and make our-selves at home in our twenty-five-thousand-pound two-room apartment. We heard Fischer start up the air compressor with a well familiar roar, and soon we were on our way.

There were two balloons on the table, one filled with incompressible water and the other with normal air. These would serve as visual evidences for us to observe the increasing pressure. The latter was rapidly shrinking in size as the air poured into the tank. Our heads were ringing with the roar of the reverberation of every little sound, from the droning din of the compressor to the very act of our breathing. The air was now burning hot from its heat of compression.

We watched the air balloon drop to the size of a plum. The other remained unaffected, staying about the size of a grapefruit. We realized that we now were about up to pressure. A moment later, the compressor stopped, followed by a clang on the telephone. Fisher informed us that we were in one hundred feet of water.

There was nothing new about the sensation of this much pressure, but it was a novel outlook to realize that this would be our home for the next few days. It was certainly not an unpleasant prospect. The pressure, as it always does, lent a pleasant feeling of intoxication that made it rather like

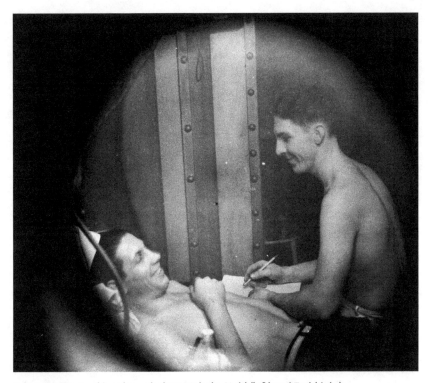

Taking it easy. Looking through the port hole. Nohl (left) and End (right).

a gay party. We talked a while in our thilly lithps, and then I lay down to loaf and start on the book that I had been wanting for so long to read.

We had each had blood drawn as we entered the tank, as a normal base for comparison. Under full pressure, we were now ready for the next sample, but there were not enough doctors to do it. Dr. End could draw mine, but I would be the only one there to take his. I had never done this before but had had it done enough times to have an idea of the procedure. I dug around in his arm before finally finding the vein, and we passed the sample out through the air lock to the waiting technician for analysis.

We decided to take them every hour for the first six hours and thereafter every two hours. By the second day, we might cut that even further depending on the laboratory reports. The difficulty was that there would be almost a four-hour delay from the time that we drew blood until the time that we received over the telephone the report as to its acidity. This was a very careful and exacting laboratory procedure and could not be done in a few minutes. There was a considerable additional loss of time in getting the blood out through the outer chamber, or airlock, of the tank. We were not particularly happy about this, but there seemed to be no way to decrease the time.

Thus, four hours after we went in, which seemed an eternity later, the call came in on the condition of our blood before we started. It seemed to be about what it should, according to Dr. End.

At five hours, the one-hour test came through. The alkalinity was a little low. At six hours, the two-hour came through. It, too, was a little low. It was apparent that it was dropping but not alarmingly so. The three-hour report at seven hours was still on the same trend—slowly dropping. Dr. End was worried.

If anything went wrong it would go wrong with both of us. He felt that it was his responsibility to be alive and healthy himself so that he could treat me in case I got into trouble. Furthermore, outside the tank, he could help in the blood analyses.

He still had two full hours of decompression ahead, assuming that he had reached saturation at three hours and that the hours from three to seven were "free."

I hated to see him go, since it would be awfully lonesome in here by myself, but I had to admit that it was a sensible thing to do. I moved into the back room, and he stayed in the front, which would now serve as the airlock. Fischer sealed the door between us by dropping the pressure a couple of pounds in End's side. In feet of water pressure, he dropped to thirty feet quickly then took stages at thirty, twenty, and ten feet, after which the doctor finally emerged from the tank. During this time, no blood samples could be passed through, since End was occupying the airlock.

As he emerged from the tank, he received the report on my four- and five-hour samples. The alkalinity was lower and apparently dropping at a greater rate than it had before. An hour later, the six-hour report came through. It was dropping rapidly. Dr. End called me. He was afraid that the curve was dropping off too rapidly and would, by now, be approaching extremely dangerous lower limits.

At eleven hours, the seven-hour report was ready. The slope of the curve had changed again. It was dropping precipitously. Continuing it on down at the same rate to where I now was, the conclusion was quite alarming.

I was theoretically dead.

My blood acid-base balance was below that at which a man could live.

He told me over the phones—me, now officially dead—what had happened. Whether I liked it or not, I was now coming out of the tank.

I didn't know whether to be brokenhearted or jubilant. Our three-day experiment was now all over except for the decompressing, provided that I could do that. This would be a ticklish problem. With my blood as it was, I would be dead without the pressure. I would not be further exposed but would then suffer from the full blunt of my low alkaline, or highly acid, blood.

There was only one answer—I had to try to come out and see what would happen. I called in the OK, and soon I heard the roar of air as Fischer opened the exhaust valve.

In reverse of what had happened during compression of the gases, the expanding air suddenly turned the tank into a cold winter day. I grabbed for a blanket to cover my warm, sweating body, and we started "down the hill" to thirty feet. Dr. End, alarmed about the condition I was in and not feeling too well himself, asked that I be decompressed on an extralong schedule—three hours instead of two. I stayed at thirty feet for about fifteen minutes, twenty feet for an hour, and ten feet for an hour and forty-five minutes. When the last stage was over, Fischer called that they were now ready to let me out, fifteen hours after I had entered.

Dr. End had been overcautious, I thought, as I saw the air balloon approaching almost indistinguishably the size of the water balloon again. I felt fine, if a little drunk from the pressure.

I was now almost out, only about four feet of water pressure—that is, two pounds per square inch—left in the tank.

Suddenly, I thought the fury of heaven had struck me. A violent pain shot through my head, arms, and legs. I let out a yell that almost loosened the rivets in the tank.

The pressure dropped to two feet of water—about that in the bottom of a bathtub, one little pound per square inch—and the pain was growing worse. It was now so intense that it was indescribable. The nearest description of which I could think was that five teams of twenty mules each were connected respectively to my two arms, two legs, and head, and they were pulling as hard as they could in opposite directions. In the icy air, I broke out into a cold sweat. I couldn't stand it.

My legs were getting worse. What on God's green earth could be worse than this?

The last pound of pressure dropped away, and suddenly the big steel door swung open. There was a confusing chatter of many voices—people were rushing in, there were flashes of light from the press cameras, there was a feeling of being suddenly far out in space. I couldn't quite be sure of what was happening. I was blinded with pain. My head was reeling. I was far away in another world, deliriously watching the people coming at me and yet not sure where I was.

Suddenly, my legs crumpled beneath me. They were paralyzed from the hips down. I automatically thrust my hands out to break my fall as I slumped to the floor. Like my adjacent spirit, I objectively watched myself and wondered what it was that made me instinctively reach out to protect myself.

The pain in my abdomen was growing in intensity, now almost seeming to turn into a paralysis. I was scared stiff. I really had the bends.

Dr. End ordered that the door be shut after he had helped me on to the cot. Fischer was building up a supply of air in his auxiliary receiver, which he turned loose into the tank a minute later. That blast of pressure was like a magic wand, running a soothing hand through my entire body. It did not completely relieve the pain or paralysis but cheered me up as I realized that it soon would help more.

The compressor roared away, and I felt the pressure building up as the air ran through my Eustachian tubes. And then, suddenly, a miracle seemed to occur. The pain disappeared throughout my entire body like magic. I felt fine. I tried to move my legs. There was no more paralysis. I stood up, walked over to the phone, and called Fischer, asking him to hold everything. My bends were gone!

It seemed advisable to stay right here at this pressure for an hour and adjust myself to it. Thoroughly accustomed to it, I could then make another attempt to get out. Fischer called back and said that I was under twenty feet of water pressure (8.8 pounds per square inch).

I never felt any better. I was in good spirits again and pleased to note that I was so close to zero on the gauge.

After the hour had passed, Fischer advised me over the phone that he was now preparing to drop the pressure and let me out.

Down it went—8—7—6—5—4—and then—BANG!

A streak of pain shot through me like lightning. My legs shook with the shock—and then disappeared. I was paralyzed again. I couldn't get over to the phone to tell them to stop but just screamed, "Hold it!!!"

Apparently, he had heard, for almost instantly the valves closed, and the noise of the escaping hissing air was replaced by the starting-up roar of the pulsating compressor.

The magic wand soon was again waved. As quick as the blink of an eye, the pain fled from my body. My legs were all right. I felt fine. Fischer called in, "Nine pounds" (twenty feet of water). I was right back where I started.

I waited another hour. He tried it again. Someplace between three and four pounds, the fury of hell hit me again. Up we went on the pressure until I was comfortable.

Dr. End came in the tank through the airlock. For the next hour, he wanted me to breathe pure oxygen. He had a mask, which he strapped over my mouth. This would decrease the nitrogen percentage in my lungs by displacement and allow that gas to more readily pass out of my blood.

An hour later, I tried it again—this time I was able to drop to three pounds before it hit me. The next time, End asked that I be put under full pressure—forty-four pounds (one hundred feet of water) for a few minutes and then try it again. There was no trouble dropping the first forty-one of these forty-four pounds, but just at that three-pound point, again, came the streak of lightning.

Next, he tried holding me at three pounds, just a shave above the point where the bends came. Fischer then dropped it by ounces, getting me down finally between two and three pounds before it came again. Up we went to fifteen pounds where it was now disappearing and then back down to try it again.

Hour after hour after hour slipped by, Fischer pressing that lower limit like a fisherman playing the slack on a swordfish. Each time he went too far and the pain hit me, he went back up again until I called the OK. The pressure at which I found relief was gradually getting less and less, dropping down to twelve, ten, and eight pounds. The lower limit was also gradually approaching the zero for which I dreamed, finally dropping below two

The pressure is up and the bends are gone. Dr. End (right) massages Nohl's still sore leg muscles.

pounds and then below one. Of consolation to me, as I wondered if I ever could get out of this mess, was the fact that each time the bends came back, it was with less severity. Finally, the paralysis completely disappeared, and it resolved itself into mere pain, even this becoming comparatively more and more bearable.

Twenty-seven hours later, as Fischer was easing down the pressure in fractions of an ounce, I heard a creak and felt for the second time the strange sensation of being connected with the external world by some sound reverberation perceived in my ears. I turned quickly. The door was open.

There was still a penetrating rheumatism-like pain shooting through my arms and legs. I could walk with a sort of shuffle. My arms were in such a state that I could move them but did not have the strength to raise them over a few inches. I could relieve the pain by going back under pressure again, but I decided to stick it out this way. Dr. End rigged up a sling for my right arm, and that was the end of the dive.

The following day, the pain had largely disappeared, having turned into more of a soreness and weakness. It was a bit difficult eating, but by the third day, I took off the sling and almost forgot all about it except for the occasional pang. A week later, it was just a memory.

I was very happy about it all. There would be no permanent ill effects whatsoever. What we had learned might well be an important key to the exploration of that great last frontier on this earth—the ocean floor.

Although we had not proven that bubbles cannot have any part in the bends, certainly it was now very clear that the explanation of the insidious disease was something far more complicated than the classical fizzing soda pop analogy would explain. If bubbles had been the entire story, after three days in the tank, I would have been no more saturated than I would have been after three hours. I would not have had the bends for my additional long exposure. The increase in blood acidity with exposure was now a proven fact.

In summary, our progress had followed this rather strange pattern: We had always previously assumed, but with considerable skepticism, that the classical explanation of the bends (the fizzing bubbles of gas dissolving under pressure) was correct.

We had selected for an artificial air the gas helium because of its low solubility in blood and tissue and the obvious advantage of involving smaller quantities to pass off as bubbles.

We had tried it, in turn, on the guinea pigs and other animals, on ourselves as human guinea pigs in the tank, and then in the new self-contained diving suits where it seemed to work far better than our greatest expectations.

At tremendous pressures, it seemed to offer the even more important advantage of allowing a man to stay conscious, and we had been able to reach 420 feet for a new world's record.

Tying together the many phenomena that the bubbles had failed to explain—the lack of toxic effect on the mind in deep water of helium and the now evident connection between blood acidity, the bends, agglutination, and the causation of these by air under pressure but not by helium—we had decided to test our theory in the long exposure dive. The suspected increase in acidity had now been indicated, and we understood why helium worked so well. With this as a start, it would now be possible to intelligently plan a procedure for its use in deep water.

In the world of science, theories are not proved by just one experiment. When dealing with the human body, there are too many highly complex or even totally unknown factors entering into any physiological experiment to make quick conclusions advisable. Thus, from this single incident, we could not and would not make any sweeping inferences. However, we did think that we had found at least another important clue as to what the bends really were and possibly another key to man's invasion of the great depths of the ocean.

Book VI

Sponges and Celluloid

33

Underwater Wonderland

Out of the dark depths of the earth flows a mighty subterranean river, filtering its rushing waters through countless miles of Florida's limestone base, then turbulently breaking through the earth's crust to gloriously bathe its crystal-clear waters in the warm light of the bright tropical sun.

A pool of turquoise blue, surrounded by green towering trees of a fertile virgin jungle. Water so deep, so clear, so mysteriously blue that callous would be the heart which would not jump at the sight of her wondrous depths.

Since this river is as clear as the mythical pool of paradise, it's no wonder that the wary Ponce de Leon triumphantly cried out that he had at last found the long-sought Fountain of Youth.

Wakulla Spring! Water of Mystery! A place appropriately named by early Indian Americans who were apparently always drawn to this heavenly pool and around whose banks centered the activities and legends of many tribes for countless generations.

This fascinating spring had been part of the great undeveloped empire of Alfred I. DuPont. It was located in that corner of Florida that the world had somehow forgotten, along the Gulf Coast at the turn of the state's characteristic panhandle.

Alfred DuPont had died, but his successor to the estate, Edward Ball, was determined to carry out DuPont's uncompleted vision. I was invited to conceive, design, and construct a number of pieces of underwater equipment that would attract interest to the spring.

The water was perfect for photography, and so we decided to build elaborate equipment to facilitate the production of professional underwater motion pictures. My first "baby" was the strangely but appropriately named Hole in the Water. It was just that: a glass- and steel-lined depression down in the water, designed for professional subsurface motion picture work. Open to the atmosphere at the surface, it was covered with a light trap so as to avoid undesirable reflections on the windows within. The well, cylindrical in shape, was electrically controlled so that the cameraman, at the touch of a finger, could turn the entire four-ton chamber to pan, or to follow his action, in the water.

The next novelty was to be a miniature underwater house, large enough to allow four or five strangers to insert their heads and shoulders for breathing and resting as long as they cared to stay down. Fed by an air compressor on the bank, the first house was placed at a depth of twenty-eight feet, a substantial swim for anyone. The second similar house was planted at ninety feet, and the third was to be at the maximum depth of Wakulla Spring, 185 feet. Although these distances seem progressively longer, our concern was with the ratio increase of the absolute pressure, which in each successive lap was to be approximately doubled.

The houses were constructed of Lucite, which not only made a fine water-resistant material but gave the occupant a magnificent view of the surface of the spring above. From down there, boats could be seen hundreds of feet away, floating lazily on the water like little dirigibles. It was always quite a surprise, with this feeling of utter isolation, to look up and see, through the pane of a glass-bottom boat, rows of round faces grinning as they peered down into the water at us.

Nohl's sketch of the glass-bottom boat as seen from underwater.

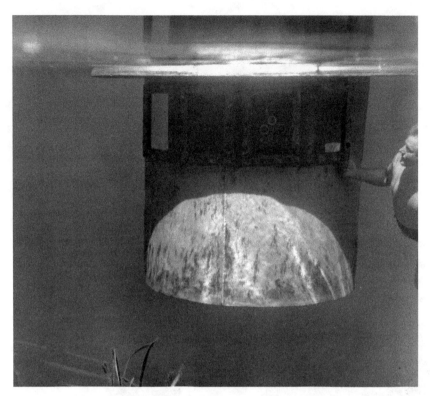

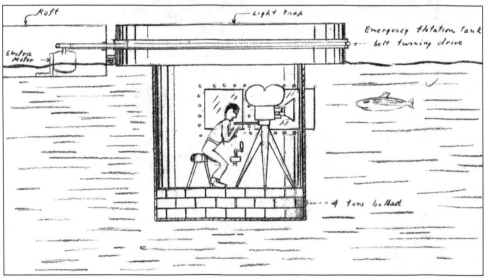

Top: The "Hole in the Water." Newt Perry peers from the outside into the submarine photographic chamber. Note the four-lens turret in shooting position.
Bottom: Nohl's sketch of the "Hole in the Water."

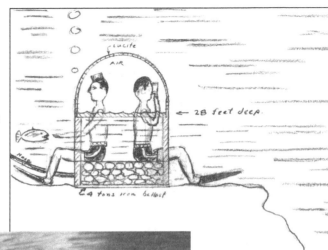

Nohl's sketch of the underwater house at twenty-eight feet deep.

Underwater house. Air is supplied through the hose (leading off to the left) so that the swimmers may stay as long as desired. Note that the two lower swimmers' heads and shoulders are in air and that the rest of their bodies are in water. Similar to the open helmet, the pressure forces the water down. COURTESY OF KATHY END

Attaching rubber fins to our feet actually increased our speed through the water several times, effectively decreasing these distances to moderate amounts for an ordinary swimmer. A flat glass mask, fitted over our eyes, clarified the vision to the extent that we sometimes wanted to pinch ourselves to be reminded that we were really in water.

On the hot Florida summer days, we would often go down to the twenty-eight-foot house and spend hour after hour there. The waters of Wakulla, coming out of the deep, cool caverns of the earth, were always at exactly the same ideal temperature, winter or summer—seventy-two degrees. It was seldom that the visiting swimmer could drop down to the first house without finding several friends vacationing there. With this as a base of operations, it was fun to take leisurely excursions out into the surrounding submarine scenery, trying to sneak up on a huge turtle, a school of surprised fish, or some of the many formidable-looking but harmless six-foot alligator gars.

Of all things, we even made a "parachute" for an "emergency" trip to the surface from the house. Resembling an umbrella in size and shape, it

Perry makes a parachute ascent to the surface. The steel chute is filled with air, giving it buoyancy for a rapid rise.

was made of steel. To bail out, it was simply necessary for the would-be jumper to reach out of the window and hold his chute over the stream of bubbles exhausting upward toward the surface. The collected air would soon give it buoyancy, and all that was then necessary was to let nature take its course. At the surface, the swimmer would merely turn the chute in the water so as to spill the air and let it sink for the next jumping enthusiast.

The next novel device was intended for entry into the water—a staggered diving tower with boards at ten, twenty, and thirty feet above the surface. After we completed the huge structure with legs reaching to a twenty-foot depth, we were amazed to discover the curious illusion created by the crystal-clear water. The prospective diver, standing on the top platform, if he dared to look down before he dove, was chagrinned to find that instead of a thirty-foot drop to the surface of the water, there appeared to be a fifty-foot span to a jagged rock bottom, the water being so clear that it could not be seen. Mentally assuring himself, if he could, that it was there, he would proceed to do his dive, gracefully going into his jackknife, gainer, or midair turn. On some of these first dives, there was a disaster waiting, the water being so transparent that it was impossible for him to tell when it was time to straighten out. However crystal-clear it may have been, it was still just as hard as concrete. We finally arranged to ruffle it, so the surface could be seen for the would-be bird-men.

Next came the Glass House.

Although we were in Florida, the land of sunshine, we had long since discovered that it can get very cold in the winter. Particularly up here in the northern part of the state, it was not unusual to have weather near or below freezing temperature. However, the water was always exactly the same ideal temperature—seventy-two degrees—making possible identical swimming conditions in winter and summer. It was the going in and coming out that was so painful.

Since we were anxious to go into production regardless of the season, and since the swimmers could not stand the exposure to the air in the winter months, an almost fantastic floating house emerged as an answer. The house, built entirely of flexible glass, was always warm as an oven. The winter sun poured through the transparent roof and made it like a Turkish bath within. The prospective mermaid would enter the house in her fur

The staggered diving tower. COURTESY OF KATHY END

coat and adjourn to the dressing rooms (the only part not transparent) to slip into her bathing suit.

She would then, in Santa Claus fashion, step into the fireplace to enter her special domain via a ladder leading down into the water. Ducking under the submerged edge, she would instantly find herself free in the great spring. When she was finished with her swim or submarine

The Alligator Boat arrives at the Glass House where Newt Perry (far right) greets his visitors. The swimmers will emerge from the water through a trapdoor inside of the warmed house, thus never exposing their wet bodies to the cold air.

performance, she merely had to reverse the procedure, returning up into the fireplace to enter the warm house without ever exposing her body to the cold air.

The Alligator Boat was another novel device we worked out for transportation through the trackless wastes of the Wakulla swamps. She had so little draught that she could almost navigate on a heavy dew.

We were now ready to go to work—and before we knew it, the motion pictures producers were waiting in line to start their cameras rolling.

Grantland Rice Sportlight was the first to come in to film *Shooting Mermaids* (released by Paramount). This was a fantasy wound about a

Underwater picnic. The men build the fire as the women unpack the lunch.

mythical underwater picnic and presented what seemed to us to be a very funny comedy.

The picnic was enacted by a group of our local expert underwater swimmers, Bunny Lowry and Elsie Davis taking the female leads and Newt Perry, one of the world's finest underwater swimmers, doing the star stunting of the show. Rod Warren and Russ Erwin, Grantland Rice's veteran cinematographers, did the honors at the cameras.

There were a few technical problems involved in staging a picnic underwater. First of all, we needed a campfire. To create the effect, we buried a hose in the sand, which led to the fireplace. As Joe lit the iron-weighted logs with a "match," a blast of compressed air was released into the hose and the fire immediately started roaring, giving off a violent stream of bubbles that nicely simulated the desired effect.

The "coffee," after being duly heated on the above fire, presented quite a problem, since we were anxious to show it being poured into the cups. There seemed to be no liquid we could find that would pour downward under the water. Finally, we filled the coffee pot with fine lead buckshot, which did the trick perfectly, the camera a few feet away recording it as a flowing black stream.

There were to be two kinds of sandwiches—those to look at and those to eat. The bread for the former was sawed out of half-inch pine board nailed together with a similar pine "meat" filling. The latter were real sandwiches but did not stand up very long in the water, although Newt and Bunny were actually able to eat them down there. The apples, celery, and bananas were also real and made for more practical eating than the soggy sandwiches.

A passing fisherman, as our script went, had made a catch, but before he could hoist it up, one of the boys underwater removed it from the hook for his fried fish dinner, politely replacing it with a note saying "Thanks." Seeing this, after having a moment before hooked a fresh cucumber, the bewildered man in the boat decided that it was time for another drink.

After the meal was over, one of the boys decided to enjoy his after-dinner smoke. Taking a deep drag on the white-painted iron pipe cigarette, he exhaled a startlingly realistic "cloud of smoke." It was evaporated milk. Just before the scene had been taken, he had stuck his head above water and taken into his mouth the best part of a glass of the "canned cow." He ducked down, the cameras started, he went through the motions of lighting up, and then suavely and slowly blew the deceptive liquid out into the water.

As the finale, the inebriated fisherman, seeing the fuming clouds of smoke (dry ice, immersed under the surface) emerging from the water, picked up his bottle to examine *what* he had been drinking. He tossed it away in disgust and started frantically to row for terra firma. It was time to get out of that place.

Our next bit job was the production of the underwater and jungle "location" scenes from the Metro-Goldwyn-Mayer picture *Tarzan's Secret Treasure*.

Practically a trainload of elephants, African animals, and jungle properties arrived one day. Soon followed the cast—Johnny Weissmuller as Tarzan, Johnny Sheffield as Boy, and the inimitable Cheeta as Cheeta, the chimpanzee that would embarrass a lot of humans in intelligence. Maureen O'Sullivan, who would play the part of Jane, didn't come, since she was not particularly talented in underwater swimming. Her parts could be filled by one of our girls as a double, her close-ups to be filled in later in Hollywood.

The story, as is customary in the Tarzan series, was to take place in the heart of Africa. A tribe of cannibals was to capture Tarzan's loved ones,

Johnny Weismuller (background) and Johnny Sheffield (foreground) at Wakulla for the shooting of *Tarzan's Secret Treasure*. COURTESY OF KATHY END

and he was to thwart them as they sped away in their dugout canoes. He would swim underwater to them, come up beneath them, and then dump their boats by reaching his arm up and out of the water and pulling down on the gunwales.

Weissmuller's underwater swimming was a sight to behold. From the Hole in the Water, we could watch him as he slipped beneath the surface for the cameras and wrestled with the swimming elephants, alligators, and huge turtles, his long unshorn hair streaming out behind him.

Next, along came Universal. Their veteran cameraman, Floyd Traynham, wanted to shoot a story of a dive down into the deepest depths of Wakulla Spring in a full diving suit and then to film a "blow-up" from the bottom.

We had a huge four-cylinder air compressor taking the full outlet of an eighty-five-horsepower Ford V-8 engine. With this and a 150-pound hunk of iron, I figured that I would be able to jump overboard and reach the bottom in less than a minute. I was not planning to wear any shoes and just part of the customary lead belt so that I could "blow" to the surface with the speed of an express train. Coming up out of 185 feet of water would ordinarily require a long, slow decompression, but in this case, I would be down there for just a few seconds. I did not think that there would be time for absorption of any nitrogen into my blood and tissues and that I could do it without danger of the bends.

I grabbed on to the weight and jumped overboard, the 150 pounds in the iron more than offsetting that which I had left behind in lead shoes and belt. My ears were in perfect condition from daily dives. It seemed as if I could almost feel a gale blowing through my Eustachian tubes as I dropped like a rock. The big compressor was pouring air in, and so I had no worry about a squeeze. Reaching the bottom fifty seconds later, I looked up and saw the glass-bottom boat from which Traynham was shooting, hanging up there like a tiny toy. I could not linger here more than a few seconds, and so I immediately bent over and grasped the iron weight now settled down into the ooze and then tried to get my feet up.

The instant they passed the horizontal, a huge blurb of air ran back into them, and I turned upside down with a snap. I was still hanging on to the iron so could stay on the bottom as long as I chose.

The only thing that I had been worried about on this intentional "blow-up" was that I might come up under the overhead boat. If I did, I would probably go right through the glass tray.

As I had looked up at the tremendous area of the surface of the spring, and also at that tiny little "toy boat" up there, it seemed as if the chance of hitting it was very slim. It would be absolutely impossible to look upward as I went, to see where I was going. I just flexed my legs a little so that if I did hit the boat, I wouldn't break them.

I had my exhaust valve wide open and was now all set. Letting go of the iron, I started upward like a projectile. The air in the suit was expanding tremendously, which, plus the full outlet of the compressor, had my diving dress distended like a big rubber balloon. I was racing through the water at a terrific speed.

And then, only a few seconds later, it seemed, I landed in what seemed to be a soft feather mattress. I hit the surface and practically jumped right out of the water. After bobbing up and down a few times, I came to rest; and then a few minutes later I felt a lasso falling around my upward-protruding legs and was shortly towed in.

Left: Nohl cleaning the windows of his diving bell. A diver went down periodically to wipe the marine scum from the glass. Nohl designed and built this bell and operated it as a tourist attraction at Silver Springs, Florida, for a number of years starting in late 1939. Also designed for underwater cinematography, the bell weighed forty thousand pounds and carried twelve passengers.

Right: An artist's conception of the bell. Note the elevator, emergency ladder, internally operated emergency cable release, and microphone for communication to surface.

Traynham was a little pale. "What's the matter," I asked him.

"You were aimed right at the camera," he explained. "I didn't know what to do—keep shooting or grab my camera and run." He had kept on shooting—and by a near-surface current deflection, I had veered off just in time. "What do you say," said Floyd. "Let's go and get a little drunk."

With all of these projects, business was wonderful. I was making good money. The work certainly was interesting. However, I was beginning to long for the deep open sea.

34

SPONGE DIVING

Another young man had now come in with me, expedition-crazy Jimmy Lockwood, of Racine, Wisconsin. Jimmy was a graduate paleontologist of Beloit College and had never done anything but go on expeditions. They had not been world-shaking ventures but had taken Jimmy's restless feet into many strange places. Jimmy was a yachtsman and had sought no less exciting adventures in his own vessels. He had been haunting, hounding, and driving me crazy for so long that I could not help but realize that here was a man who had what we needed. He was a slightly built little fellow, growing prematurely bald and possessing a high-pitched voice that was never turned off, but he was sound and was made of gold. Jimmy wanted to join me on the forthcoming expedition.

I had decided to go into the sponge-diving business. On the forthcoming trip, we were going to make a study of the entire game, from the sponges growing on the deep offshore reefs of the Gulf to the marked product. Tarpon Springs, Florida, was the great American sponge port, and it was here that we would learn it.

My lecture tours had thus far been very successful. These would continue to be a fine source of extra revenue, but they could never finance the costly expeditions we had scheduled ahead. Expeditions would require fine, sturdy, offshore vessels fitted with everything in seagoing and diving equipment. I had well learned the many thousands of dollars to which such equipment will run and, after its purchase, the considerable costs of keeping it in condition.

There were certainly many wrecks to be salvaged, but they were all gambles to a greater or lesser extent. There were many alluring expedition opportunities ahead, but they too all were uncertain as to financial return.

There was only one business that I knew of which would offer the possibility of keeping constantly at work such expedition equipment and the skilled men whom we would need to operate it—sponge diving!

Arriving at Tarpon Springs was like stepping into Old World Greece.

We walked down the docks where perhaps a hundred of the diving boats were berthed. Crowded in like sardines in a can, they extended into the bayou three and four deep, their dazzling blues, reds, and greens and yellows presenting a riotous carnival of color; their owner's tastes certainly were not inhibited when the paintbrush was applied.

The countless main and mizzen masts appeared almost like a heavy thicket, intertwined with the vast maze of rigging that supported each. Looking upward, it was apparent that at the head of almost every one of these spars was a cross or an effigy of their typical Greek boats.

The stink of dying sponge was everywhere. The docks were crowded with heavily bearded Greeks, speaking quickly in their native language as they went about their various jobs.

For several years, a northern syndicate had been trying to get me to tie in with them on a large Florida west coast development plan. A substantial part of this proposal had included a surprise coup d'état on the Greek sponge industry and the fabulous fortunes that were supposedly being made there.

This industry, the syndicate stated, was under unfair monopolistic control. Their plan would break that monopoly and give, for the first time, a fair deal to the men who were daily risking their lives out in the deep waters of the Gulf. From the figures that the syndicate had gathered, that feudal control certainly needed cracking, but it was also apparent that they were not planning to do it for purely high and lofty philanthropic and sociological motives.

Thus, they were in touch with and had the sympathy of a substantial bloc of the leaders of the sponge industry. Although I had signed no contracts with the syndicate, I was interested in it if it could do what its charter stated; they were interested in my participation in the proposed forthcoming revolution and the introduction of our new types of more efficient diving suits.

Load of sponge.

I thus had high credentials and immediately could enjoy meeting the big shots in the industry and learning it from those men who had for so long been prominent in it.

The Greeks had been diving for sponges for thousands of years. Since almost biblical days, they had plunged down into the deep waters of the Aegean Sea, searching for this lowly animal of commerce. In the old country, they had braved the depths with knives in their teeth. To sink rapidly to the bottom, a diver would cling to a huge rock. With it snugly in his arms, he would jump overboard. A rope was fastened to the rock (but not to the diver) so that the rock could be retrieved. The diver could also be recovered with the valuable rock if he hung to the line after giving the pull-me-up signal. At the depths at which they worked, a diver was completely dependent on being hauled back, for he would be far too deep to swim to the surface. He would wait until the last possible second to give the signal, utilizing the brief stay on the bottom to its utmost.

These men to whom I was talking knew. They had started as skin divers in the Aegean Sea. "Hundred—hundred-twenty—hundred-forty feet. Nothing! Hold the wind. Three—three 'n a half—four minutes. Nothing!" they would say with violent gestures.

About sixty years ago, a group of the more enterprising of these men had left the Aegean Sea, their native islands, and Greece, in search of their fortunes in that magic New World where they'd heard everyone is rich and the gold runs like the tide through the Dodecanese Islands. Their reefs, after centuries of diving, were almost devoid of sponge, and their divers were going deeper and deeper into more and more treacherous waters. In that fabulous New World, they figured, there must also be fabulous new beds of sponge.

Strangely enough, there were. In the blue waters of the Gulf of Mexico, off of Florida's west coast, they found virgin reefs of fine wool sponge worth almost their weight in gold in the markets of the United States. They sent back to their native islands for their wives and their relatives and prepared to introduce sponge diving to the western hemisphere.

At the then-tiny village of Tarpon Springs, Florida, they settled. Only twenty miles from industrial Tampa and forty miles from Florida's famous sunshine city, St. Petersburg, they started to build their own community, as utterly divorced from the United States as if they were on another planet.

Their boat builders started anew to build the fleet that would take them out into those treacherous offshore waters—boats just as their fathers and their grandfathers had built for thousands of years. The new fleet had gasoline engines in them, but other than that, they were practically indistinguishable from the models of a century ago or of a thousand years ago.

The Gulf was too cold for skin diving. Now there was available to them an ingenious new device that allowed them to remain underwater for long periods at a time—a diving suit.

The years went on, and out of the early readjustment to the new environment, there emerged a familiar diversion within this closed industry: those who were to control and those who were to be controlled.

The latter were to be the owners and captains of the individual boats. The former were to be the financiers—the men who, through the control of the money, were to be ruthless feudal lords in this quaint colony. The cycle was simple and had repeated itself over and over again.

A captain, desiring to go to sea for a cargo of sponge, would require a substantial amount of capital. The trips average about three months apiece, and during this time, a captain would have many mouths to feed that couldn't wait ninety days for the sale of the catch. His wife and children would need considerable cash with which to buy their groceries and maintain the households. Likewise, the divers and deckhands also had families, and the captain needed to provide a cash advance to each of them to leave while they were gone. On the boat, hundreds of pounds of dried, canned, and preserved foods needed to be stocked in the forepeak to feed the hungry bellies of hardworking men. Running almost continuously every minute of daylight would be a thirsty engine, which would drink up thousands of gallons of gasoline or fuel oil.

The financiers graciously provided the captains with their every need. They could have all they wanted of anything that they wanted. All they had to do was to sign a little paper—a lien on the forthcoming catch. No prices were quoted on their selections—they could just help themselves.

Three months later, the captains staggered back to port with a rich load of fine wool sponge. At the great auction at Tarpon Springs, the catch was sold to the highest bidder. The money did not go to the captain, however—his only claim was to that which was left after the expenses were taken out.

The great calculation started. Perhaps the gasoline would be sixty cents a gallon. Perhaps the milk would be thirty cents a can. The eggs might be $1.50 a dozen. There were the many items of imported Greek foods, on which their diet was largely based—these prices were even higher. There were the interest charges on the money they borrowed and the goods advanced that might be 25 percent of the amount loaned.

These prices were always different, but after it was all figured out, there was virtually nothing left for the captain. He could go out and get drunk for a few days, perhaps, and then he was right back where he started—broke. He would then have to get ready to go out again; thus would start another cycle.

I learned from the men who bought and sold the sponge the answer to every possible question of which I could think as to marketing, treating, shipping, price fluctuations, market manipulation, grading, etc. It was obvious that there was certainly a substantial amount of money involved in supplying North America with its sponge.

The part in which I was most interested was now coming—harvesting the sponge. It was arranged that I signed on as diver on one of the sponge boats—strangely enough, the only boat that I had seen whose name I could pronounce, the *St. Phillip*.

I was taken down to the dock and introduced to her captain and crew. The captain grinned, his white pearly teeth glistening in contrast to his weather-beaten, begrizzled face. He pointed at himself and blurted, "Pete," then pointed at his first diver and added, "Angelo."

I understood and repeated, pointing at each respectively, "Pete" and "Angelo," and then at myself and said, "Max." They both grinned from ear to ear and started talking a blue streak, indicating, as their interpreter told me, that all was understood and we would be very good friends.

We shoved off before daybreak the next morning, wending our way out through the beautiful bayous, so characteristic of Florida's west coast. Soon we felt the swell beneath our feet, and the little *St. Phillip* boldly plunged into the rolling seas as her sturdy hull had done for many years in the past.

Angelo and Pete had very little to say to anyone once the boat left the dock. Everyone had his job to do. There was not a single command spoken during the entire voyage. If there was a duty to be done, it was unnecessary for Captain Pete to mention or even suggest it—someone saw the need and instantly did it.

I had stayed at the hotel the night before and had gone directly down to the boat in my seagoing clothes, planning to have breakfast on the boat. Our cook was fussing away with his charcoal stove on deck, getting a good set of coals burning, and breakfast was not ready until we were well offshore.

The breakfast, usually a substantial meal for me, was quite a shock— nothing but coffee! I don't ordinarily drink coffee, but seeing that that was the only thing on the morning menu, I decided I had better momentarily change my policy.

I took one sip and spit it out. It was strong enough to burn the skin off my tongue. There was no cream or milk or sugar—just that wicked black brew, the consistency and appearance of a soggy solution of black mud.

I jumped at whatever odd jobs I could see, pumping the bilge, coiling lines, but finally everything was in order, and I just curled up with my gnawing stomach and watched the water scudding by our hull.

The *St. Phillip* was practically identical to every other boat in the diving fleet. As a yacht, she would be classified as an auxiliary yawl or probably even more suitably, a yawl rigged motor sailor. In practice, her sails were a very secondary part of her running gear, a small mizzen or jigger being bent for only occasional use, particularly at night to steady her at anchor by keeping her bow in the wind. A jib was also rigged, which balanced her jigger so that she could sail with the two in an emergency or use them as helpers with a good breeze.

A breeze came up midmorning and Captain Pete broke out her two patches of canvas. The increase in speed was immediately apparent, as well as the pleasant steadying affect as we heeled to leeward.

Although she carried a large mainmast, there was, as on the other boats, no sail for it. Its function seemed to be to hold the Greek cross high to heavenward, to provide support for the jib stays, and to hold up the typical overhead framework. This was used for a sun canopy or for drying the sponge.

Noon came and went, and I had as yet seen no signs of my much-needed lunch. Our cook was fiddling with numerous other things in no way related to his prime job of providing delectable food for his crew. Finally, I could stand it no longer and went to him and tried to say in my slowest English that I was hungry.

"No savvy!" was all that he could say.

I then rubbed my belly painstakingly, pointed at my mouth, and then went through the motions of eating.

He grinned and understood, waddling over to the stove, which was still glowing. Leaning down into the cook box, he worked over something for a moment then emerged with a steaming cup and handed it to me.

It was more of the black mud. I didn't have to taste it to tell this time. I said "No!" as I pointed at the bitter brew, then went through the sign motions again of being hungry.

He understood, grinned, and took the coffee. He poured it back in the pot where it could stew the rest of the afternoon and get a little stronger. He then went below and momentarily appeared with a huge loaf of Greek bread, a jug of olive oil, and a huge hunk of the native goat cheese. The cheese, worth $1.50 a pound at standard retail prices, was hardly tastier to me than compressed cottage cheese, but now no delicacy designed for a king could have touched the spot any better. I didn't know just what the olive oil was for and made myself content with the bread and cheese.

I was learning one of their superstitions—it is bad to eat before diving.

Why, then, if this was based on physiological fact, was there no food provided for the cook, the lifeline tender, or the engineer who might live their entire lives without getting their heads underwater? Perhaps it is so dangerous that just to have the men on deck eat will result in certain disaster for the man on the bottom. I had always found, on the contrary, that diving was hard work, and the best way I could prepare for it was by fueling myself with a substantial meal.

About the middle of the afternoon, Captain Pete brought out the lead and started running a series of soundings. He stood far forward, clinging to the forestay, and stared down into the water, apparently studying it for changes in color. This would indicate the type of bottom beneath the boat. As he pulled up the lead each time, he would turn it over to study the fragments of bottom that might have clung to the chunk of soft yellow soap he had inserted into its hollow.

Time and time again, he repeated this process with never a comment. Finally, he turned as he completed the inspection and caught Angelo's eye at the wheel. He raised his arm. No words were necessary. Angelo threw in the clutch, the cook threw overboard a flag buoy, and we were ready to dive.

I had hoped that one of them would take the first dip so that I could observe the details of their own particular system of diving, but they apparently were going to give me the harshest possible examination. I was to make the first dive.

If I were not a diver—and I certainly was not a Greek—they wanted to know it right away. If I was a professional, even though this gear was somewhat different in detail from the regulation commercial equipment, I would be familiar enough with its basic principles to adapt myself to it.

I had studied over the equipment on the trip out and found that it was different in many details from the regulation navy suits. The windows were bigger, there were no protecting bars, the front faceplate did not unscrew, there was no hand-operated or automatic exhaust valve but only that actuated by a nod of the head on the inside, the dress was considerably lighter in weight, and there were quite a few other variations from our own gear.

Those men had sweated oil of garlic for so long into the dress that the odor was strong enough to curl my hair, I noted as I donned the apparatus. In addition, there was the usual stench of perspiration, rubber, and

Captain Pete (left) dresses Nohl (right) on board the *St. Phillip* at sea.

compressor oil fumes. The compressor was now chugging away with a slow, steady monotony, and I was ready to jump overboard. Firmly gripping the handle of the sponge fork in my right hand and the handle of the sponge basket in my left, I leaped off the bow and started down.

I resolved to show these Greeks that I could handle myself in gear and leaned as hard as I could on the exhaust valve. I did not have as heavy a suit as even the lightest of my own, but I found that I could settle to the bottom in almost a minute. I didn't know the depth, which had officially been announced in Greek but, by watching Captain Pete as he ran his soundings, estimated that it must be about fifteen fathoms, or ninety feet.

The lead had not lied. I was on a dazzlingly beautiful coral reef. The water was a breathtakingly brilliant blue, contrasting with the many yellows, oranges, greens, purples, and reds of the rainbow-colored reef. Massive convoluted pieces of descriptively named brain coral were lying everywhere. Towering structures of similarly appropriately named stag-horn coral loomed up from every side. Brilliant purple- and orange-colored sea fans and plumes waved lazily to and fro with the movement of the waves almost a hundred feet above. Under my feet was a huge starfish almost a foot across. A few stops ahead were half a dozen sea urchins, black little fellows bristling with hundreds of poisonous quills like a living pin-cushion of lethally charged hypodermic needles. I had long since learned never to touch these fellows.

This was one of the most beautiful sights that God had ever created—a coral reef teeming with life, each species fighting in its own strange way for the very right to exist, to feed, and to propagate its own kind. It was only on such a reef that a sponge, one of the lowest of all animals, could live, but finding a reef did not necessarily mean finding sponge.

Little has been learned about the means of propagation or means of sustenance of this animal, but it is known that the sponge, seeking a place on which their sperm may grow, will choose something solid. They will never settle on sand, shell, pebble, or mud bottoms, and thus, for practical purposes in the Gulf waters, coral is the only solid substance to which they can cling.

Pete had found this reef, and I only hoped that it would be a good one for sponge. If it weren't, I could go up and say "No sponge," but they would not be sure whether I knew a sponge when I saw one and whether my judgment was correct. Perhaps just one out of ten reefs are sponge bearers, and I crossed my fingers that this would be one.

I started crawling along over the jagged coral. At each turn, I came across a few more goggly-eyed fish, who would stare curiously at my mon-strous form for a moment then dart away to more certain safety.

Looking out of my left window, there, only a foot away, was an enor-mous sponge. As I turned and approached the animal, it seemed to be the biggest one I had ever seen—it must have been thirty inches across. I regripped my three-pronged fork, getting ready to go to work, and then dug the tines in as low on its base as I could. Prying back, I found that my hook was tearing itself through the sponge. It seemed to be rotten and crumbly.

And then I remembered. This was not a sponge. Perhaps it was bio-logically, but it wasn't commercially. It was one of those things that they had warned me about—the loggerhead. I started tearing at it from all sides and found that in every appearance it fully resembled a sponge; yet, in strength, it was more like a soggy newspaper. I studied it to make sure that I wouldn't make this mistake again.

"Watch for the eyes," the Greeks had told me. From the pieces I had studied in Tarpon Springs, I had to admit that the sponge's ventricles actually looked like eyes. These were the mouths through which the animals took seawater to pass through their bodies, from which they absorbed nourishment, and into which they passed their waste products.

And then, there he was, just ahead of me, staring at me with his big black eyes—my first sponge. He was a dandy, about seven inches in diameter, a first-quality deep-water sheepswool.

For forty minutes, I crawled around through the jagged formations of the reef, pulling one after another. I thanked my lucky stars that I had hit such a rich reef. They were largely the almost priceless "wools," although I had found a few yellow, wire, and grass sponge varieties, too.

The yellow is much like the wool in shape but, after cleaning, is more of a yellow color. It is only worth a fraction of the value of the wool, since it does not have the softness, toughness, or absorbent qualities of the former, although it takes a second place to it. The yellows are usually the cheaper five- and ten-cent store or drugstore sponges. The wire sponge is somewhat similar in shape and color to the wool, also, but is characterized by a fine wiry-looking exterior and is inferior in every way to the above two. The grass sponge is the other common Gulf variety and is probably the cheapest and poorest in quality of them all. However, it is so plentiful and when found comes in such enormous quantity and price that many divers will pull them. The latter three kinds are known as "expense sponge"—if luck is very bad, they will pay for the food and fuel and prevent the book from going into the hole, but the profit for which the divers are always looking is in the wool. Some boats will touch nothing but wool, assuming that every minute spent pulling the cheaper grades is just that much time lost from looking for the more expensive ones.

The Greeks have a word for every part of their business. The process in which I was now engaged was known as "catching sponge." Visitors to Tarpon Springs, many learning for the first time that the sponge of commerce is the skeleton of an animal that lives on the bottom of the sea and hearing the expression "catching sponge," arrive at a rather startling picture of a diver chasing them around in a deep-sea game of tag. The thought of such a pursuit is rather amusing, since the sponge anchors itself once and for all so firmly to the reef that it is quite a sizable task to get him off. With the diver's fork, the tough structures of the sponge must be torn loose by shear strength from their tenacious grip on the rock below, requiring in most cases a number of minutes of hard labor.

The sponge, once loose, is placed in the net bag on the diver's left wrist. This will hold, if crowded, a dozen five-inch wools or one big grass, if you can get it in. When loaded, or a few minutes before, in anticipation, the diver gives a pull signal on his lifeline. The tender up on deck immediately ties an empty basket to the lifeline at the point where he is holding it and then signals the diver. The latter pulls down his own lifeline, coiling it on the bottom until he gets the empty net. Untying the empty, he fastens in the same place the full basket, signals "haul away," and goes back to work, the tender then simply pulling the sponge. I was very impressed with the simple efficiency of this system, with only a few seconds of the diver's time being lost.

I finally received my "come up" signal and prepared to leave the bottom. This was the part of the dive of which I was most skeptical, because it was in sharp contradiction to every principle of diving that I had ever known.

The Greeks didn't decompress. I was horrified when I first learned this, after having spent so many years in the study of the effects of excessively rapid ascents and their resultant disease, the bends. I had picked up quite a few minor cases of the bends in my past years and had a tremendous personal respect for them, besides that which I had developed by intensive study of the disease in others.

I had at first concluded that I would never want to take a chance diving with the Greeks, but after much study, I had now decided that they had worked out an effective means of avoiding the bends with no understanding of the disease. It was an empirical formula, which had been developed from the results of generations of diving. The cemeteries were filled with the bodies of Greek divers who had lost their lives in token to this mysterious disease but from whose experience they had learned what they now knew.

The Greeks had found out that if a diver would stay down forty-five minutes at a given depth, he would come up dead. After a thirty-minute dive, he might just get a severe paralysis or go blind. After twenty-five minutes, he would show only mild symptoms. After twenty minutes, he would usually be all right.

Therefore, twenty minutes was the right time to spend on the bottom.

Every so often, a man would spend only his approved twenty minutes on the bottom, come up, and fall over dead. They gradually learned that if the diver was excessively tired, had been drinking heavily, or had worked excessively hard on deck, the risk was inordinately high. Divers are thus now forbidden on all boats to do anything—they mustn't strain themselves. The deckhands will wait on them. They are like idle kings.

For every depth, they had learned their time limit in the very hard empirical way—and then, when the period was over, they went up to the surface like a skyrocket.

I shot up with my fingers crossed. In the Greek system, the diver wasn't pulled up—he merely inflated his suit and floated up. I appeared at the surface about four hundred feet away from the *St. Phillip*, floating high enough so that my window was out of water. I could see the boat through the dripping pane, pitching to and fro as she eased about in her course to go over and fish me out.

I had no demonstrable case of the bends, although I could feel the bare start of some of those well familiar borderline symptoms. The Greeks certainly were not playing the game with much margin of safety.

As they took the hat off my head, Captain Pete and Angelo were grinning. At the same time, they both started talking a blue streak of Greek, not a single word of which I could understand, but which I liked to assume meant that they were pleased with the catch I had sent up. "No savvy," I said.

Captain Pete went over to the sponges, picked up one of them and brought it over, grinned, and took my hand and shook it heartily.

I was very happy. I had made the grade with them in their own game. They were my friends.

Captain Pete looked at the sun and decided to take a dip himself—there would be just time for a short one. I watched Angelo as he dressed his master and, as I saw him slip overboard and disappear in the deep blue water, realized that I had done things just about the same way they did them themselves. My preliminary instructions before we sailed had covered the entire story.

Pete came up just as the sun, a magnificent bloated orange ball, was dropping beneath the water.

Our cook had been hard at work during the last dive in preparation of the daily meal. I had watched him as he opened can after can of strange-shaped farinaceous substances and dumped them into a pot. Over this went enough olive oil to make a hippopotamus bilious and then the antici-pated enormous portion of garlic. In any other place in the world, I would have wondered what type of man could pass that mixture down his gullet, but at this point, I could have eaten the iron pot, too, and enjoyed it.

By the time Captain Pete was out of the gear, it was totally dark. The cook had lit a gasoline lantern, which seemed to me to be quite a bow to civilization, and we all gathered on the after deck for chow.

As a guest, I was handed my meal first, a huge bowl of the mysterious mixture. There was no spoon or fork with which to convey it to my mouth, nor did I have any idea of how to ask for one. I just set it down and waited to see what they were going to do. Pete received his portion next, as master of the vessel, and he, too, set his on the deck.

I turned around to see what the cook was going to do next and was startled by a terrific hissing noise behind me. I turned around and there was Captain Pete, flat on his belly on deck, with his face plunged into the bowl, sucking like a starved pig.

Angelo followed suit, coming up for air periodically, his mustache drip-ping with the oily stew. One by one, they all dove in.

When in Greece, do as the Greeks do. It was the only way to eat the stuff.

The slumgullion, the best name that I could think of for the stew, wasn't bad at all. I could taste garlic in my mouth for the entire rest of the trip, adding a fresh supply every day, but it didn't matter much out here.

They all sat around and talked for a while, breaking the almost perfect silence of the day. Everyone talked at the same time as fast as he could go, and it sounded like six runaway Victrolas. It didn't matter much now whether I knew any Greek or not—I joined the little group and, just for the foolish fun of it, started talking myself as fast as I could. It made no differ-ence what I said; the only thing that was important was to talk loud and fast. I repeated, as dramatically as I could, Lincoln's Gettysburg Address and the Boy Scout Code, which I had never forgotten, and then I started ad-libbing a rapid-fire description of the scene out here on deck, just as if I were giving an on-the-spot radio broadcast. I found that it was fun, certainly much more so than sitting there by my isolated self, a mere American.

Captain Pete finally yawned, yawned again, and then got up and disappeared below deck. Within two minutes, the roar of his snore shook the boat. Angelo went next and, in turn, was followed by the rest of the crew.

I had to be the last one down, since there was no other way for me to learn which bunk was mine.

I had barely ducked my head below the companion hatch before I realized that I was not going to sleep down there tonight. The stench of perspiration, mold, garlic, gasoline, and rotting sponge was enough to sprout hair on a bald man. It was cool and sweet up on deck and certainly no harder than my so-called bunk.

Grabbing the blankets, almost as stiff as a phonograph record with accumulated grime, I went up topsides, assuring myself in the darkness that I didn't know how dirty those blankets were. I curled up in the alleyway and soon disappeared into a world of dreams.

However, my dreams were soon rudely interrupted. It was raining and a cold stream of water was running in the gutter beneath me. My blanket was soaked, and as I fully awoke, I realized that my delightful night on deck was now at an end. Stench or no stench, I would have to learn to like it below.

The cabin was not what would be called overspacious. I could not stand up by a good foot. Built-in tiers to fit the curve of the hull were the bunks, from which emerged world-shaking snores.

On any deep-sea boat, these would correspond to bunks. The Greeks had their word for them, but to me they were shelves. Attempting to crawl into mine, I found that any adjective describing it as small would be an understatement. My six feet three inches were at least a foot in excess of the bunk length, which meant doubling up in a Z-shaped figure to slide in.

After a few minutes on the hard boards of the bunk bottom, my lower side was ready to turn over and be the upper side for a while. However, the Z could not turn over, my bent knees striking the upper bunk in less than an eighth of a turn. I thus could not under any conceivable conditions lie on my belly or on my back. The only way that I could turn was to get out of the bunk completely, stand up on the cabin floor, and then start anew, this time backing into the narrow space.

I finally fell asleep, the roar of the seas and the roll of the boat acting as the finest soporifics in the world. Fortunately, it didn't rain every night, and I usually could enjoy the most satisfying experience of sleeping on an open deck under the stars, fanned by the sweet cool breeze sweeping over the great expanse of offshore Gulf and lulled by the lapping waves against our sturdy hull.

It was a good trip. The catch was bountiful and of first-quality sheeps-wool. It seemed as if I had passed my apprenticeship. I was ready to go into the business.

35

TIBURÓN

I never forgave myself for not learning Spanish.

For two years, I had sweated blood over Latin in high school and then had switched to German for another two years. I had never found much opportunity to put either to practical usage except to utter the customary "Prosit" with the hoisted stein of beer.

In Mexico, a few years ago, armed with *Learn to Speak Spanish in Five Minutes* (although the title was slightly overrated), I had found that I could familiarize myself with enough stray words so as not to miss any meals or other important functions.

We were now down in the Republica Dominica, that colorful little country that shared the West India island of Hispaniola with her sister republic of Haiti. This was the land that had probably seen as much blood-shed, revolution, poverty, and prosperity as any other spot on the earth. Of interest to us at the moment, however, was the fact that off the palm-fringed shores of this island, in the clear blue tropical waters, lay some of the most fabulous treasure trove the world had ever known, the fight for which had spelled the rise and fall of great nations in bygone eras.

Jagged coral cliffs rose sheerly almost to the surface out of the great ocean depths, disastrously deceiving the wariest of navigators. Through-out century after century, during the days when this island bordered the great galleon trade routes, the bared fangs of these cliffs hungrily awaited those delicate craft carrying Incan gold from South America to the greedy vassals of England and Spain. The ocean floors off these coasts were veri-tably carpeted with the rotting hulks of mighty merchant vessels, reckless pirate raiders, and grim men-of-war alike, and with the moss-covered skeletons of the adventurous lot who chose to deal in that yellow stuff—stuff which causes men to kill or to be killed in the mad lust for gold.

I had no delusions about the chance of retrieving any part of this fantastic treasure. True, in just one mighty storm, sixteen of those great high-pooped galleons had been driven on the cruel coral crags of the reefs just north of this island, and those shoals were loaded with staggering tons of pure gold.

I was too well aware of what even a few years of submersion can do to a ship. After three hundred years, the chance of salvage could be nothing more than just that—chance. That was not what I had come for. Armed with thousands of feet of color film, a battery of underwater cameras, and a burning desire to explore those beautiful reefs that had never before been seen by the eyes of man, I had no hope of financing my expedition with pieces of eight but merely planned to browse around through the coral, check on the wrecks and sponges, and shoot pictures of anything interesting I might run across.

A few weeks before sailing from the States, we had completed the development of a novel type of diving apparatus. For want of a better name, we dubbed it the diving lung.

It actually was somewhat like a human lung, its function exactly complementing the respiratory system of the diver.

We wore it on our backs like a knapsack, the device in air weighing twenty-two pounds. Underwater, the average weight was seven pounds, the difference being due to the buoyancy of the water.

The human body has a natural average buoyancy of approximately seven pounds. Thus, with the seven-pound weight of the lung on his back, the diver discovers that his natural buoyancy is exactly canceled, giving him what is technically known as zero buoyancy. He does not tend to float. He does not tend to sink. He just stays where he is. A ghost is the only other person who can ordinarily enjoy such three-dimensional gravity-defying freedom.

Thus, wearing the new lung, the diver finds himself free in a three-dimensional world. With a flip of his foot or a paddle of his hand, he is moving in any direction his fancy might choose in space.

Inside of the protecting metal case on the diver's back is a flexible rubber bellows. As his lungs contract for his exhalation, the bellows expand; as his lungs expand for inhalation, the bellows contract. As the diver removes the oxygen from the air for combustion in his body, the lung adds fresh oxygen. As the diver pours his waste products, carbon dioxide, into the air for exhalation from his body, the lung removes by chemical process this foul gas.

The mechanism of the lung actually is somewhat similar to that of the previously described self-contained diving suit, the necessary elements of the latter here being reduced to their simplest form.

In effect, we now found that we had a miniature apparatus, completely independent from any air hose or any other connection to the surface. The lung would fit into a small suitcase. We could stay underwater with it for three hours at a time.

It did have one serious disadvantage, however. Our bodies were not protected from the cold of the water as they were in the diving suit. We could only use the diving lung in warm water. But that was of no concern now—we were in the tropics.

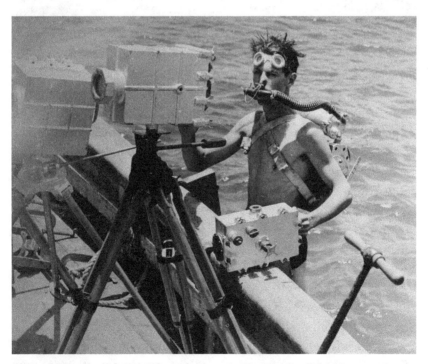

Nohl, wearing the "diving Lung," with cameras in the Dominican Republic.

The first model of Nohl's "diving lung." *MILWAUKEE JOURNAL* PHOTO

We also had another device that, with the lung, made the Buck Rogers analogy complete—our latest brainchild, the submarine pneumatic harpoon gun. This was powered by one of our standard diving cylinders containing oxygen at twenty-two hundred pounds per square inch. We had found that this would expel the harpoon with almost bullet speed. We had found that the gun, up to distances of fifty feet, was amazingly accurate in the water. To avoid the regular loss of our missiles, we frequently fastened a line to them. This was tended above from the surface ship. Thus, on a hit, our prey was readily hauled in, often to form the basis of our forthcoming lunch, sizzling in the pan as we returned from the dive. On a miss, at least we could retrieve our harpoon.

After several months of these researches and explorations in the tropical waters, I decided that there was one more objective I would like to accomplish. I had taken considerable footage of porpoises playing at the surface but had as yet not succeeded in getting any underwater shots of these fishlike air-breathing mammals. There was something about a

Nohl stalking underseas game. The harpoon gun, of his design, operated on compressed oxygen at 2,200 psi muzzle pressure and would, according to Nohl, stop any
undersea creature "in its tracks."

porpoise that I found very fascinating, and I was convinced that a motion-
picture study of them from below would be of much interest.

I had been working all summer long on my *Learn to Speak Spanish in
Five Minutes* and had accomplished enough to get along in this strange
country but was suffering from such a limited vocabulary that I could not
get over a few words ahead of the book. My conversations with the natives
were largely confined to questions about my own wording, asking for a yes
or no answer, since their speech immediately went beyond the minimum
essentials that I had learned.

One day, I ran across a local fisherman who seemed willing to labor
with me over my bad Spanish.

"Dónde halló los tiburón?" I said, intending it to be the Spanish coun-
terpart of "where can I find porpoise?"

"Tiburón?" he asked with a shudder.

"Sí, tiburón, sí!" I confirmed. I knew that the natives were seemingly superstitious about porpoises, revering and protecting them with the same devotion that the people of India might bestow upon the sacred cow.

"Yo quiero fotografiar el tiburón," I said, conveying the thought that I only wished to photograph them.

He puzzled over it for a moment then shrugged his shoulders.

"Estar aquí mañana a las tres en la tarde," he said three times, which I understood from my well-thumbed book meant, "Meet me here at three o'clock tomorrow afternoon." I tried in vain to explain that it would be better if we did it earlier in the day since I was anxious to have the brilliant noonday sun for maximum underwater light, but apparently, he didn't understand or had some reason for the late appointment that I was not able to contest, and so we agreed on three.

The following day, I arrived at the appointed hour burdened down with the heavy equipment. He looked wild-eyed at the impressively begadgeted submarine camera and diving lung, suddenly starting off with a barrage of questions, which I was totally unable to comprehend. Realizing the futility of our extended conversations, he beckoned toward his boat and we both jumped in. Soon, we were wending our way eastward along the palm-fringed coast, the almost turquoise blue water and sky beautifully setting off the massive green mountains of the fertile island.

About a half hour later, he stopped and looked around in every direction and announced, "Tiburón—aquí."

He cocked his head and blinked at the sun and added, "Por ventura diez minutos, un media hora, mas o menos," which seemed to mean "maybe ten minutes, maybe half an hour."

I couldn't understand why this was in the nature of an appointment, but my feeble requests for a slow explanation only released rapid-fire rounds of Spanish that were totally beyond my rate of interpretation. However, he repeatedly pointed down into the water and said, "Tiburón—aquí. Tiburón—aquí," and so I decided to drop overboard and see.

I sat on the stern of the small boat, dangling my feet in the warm water, and slipped the lung like a knapsack on my back. With the breathing tubes coming over my shoulders, I slipped the mouthpiece under my lips and then opened the oxygen valve. I turned to grab the underwater camera, and he suddenly seemed to sense for the first time that I was actually going down into the water. He leaped up excitedly and cried "TIBURÓN! TIBURÓN!" as he pointed frantically down into the water.

"Tiburón—sí," I answered back, supposedly conveying my realization that they were down there.

But he kept shouting, almost with tears in his eyes, "Tiburón—tiburón—tiburón!"

I took a last look at the sky, water, and shore. It was a beautiful, clear, bright day and the sun was pouring its heat onto my naked body. The water was calm with very little movement of our boat. The white sand beach was only a few hundred yards away with a long pier extending part of the way out. Diving conditions were good, and so overboard I went with a splash. I weighted the lung slightly and slowly settled down to the three-fathom bottom.

The water was reasonably clear but not of the crystal transparency that I would have liked. There was not a fish in sight, only a few lazy weeds bending to and fro with the slight surface swell. The bottom was covered with a grayish sand and, all in all, presented a rather dull landscape. I paddled along, hovering a half a fathom above the sea floor to make a little exploratory circuit of my surroundings. I could not understand what he had had in mind bringing me to this spot.

Suddenly, by some strange sense, I was aware of some massive monster moving behind my back. I wheeled around and found myself face to face with a giant tiger shark. He was closely eyeing my every movement and had come almost to a dead stop. A moment later, I saw out of the blue wall of water, the form of another, slowly easing in toward me. And then another, and another, and another.

To my left, there were three more. To my right I could see four more. In front of me, they seemed to be coming from everywhere. There must be hundreds of them—huge fellows—a thousand or twelve hundred pounds apiece. I looked upward, remembering that I was now in a three-dimensional world and saw that I was under almost a solid ceiling of shark. Their big white bellies glistened as they glided lazily along above.

I was trying to decide whether I should be very badly frightened from the grave danger I was in or whether it was only a seeming danger over which I need have no alarm. I was so scared that I couldn't seem to think.

Suddenly, there was a terrific commotion and they all seemed to spring to action. They were charging to and fro, disappearing in and out of the wall of fog that marked the limits of my hemisphere of vision. They violently swooped toward the bottom and churned up a cloud of silt on every side.

There was only one possible thing that I could do, and that was—nothing! The most foolish thing in the world would be to try to go up through the water where my white body would glisten temptingly in front of their hungry eyes. Instead, I lay down, motionlessly blending myself as much as possible into the light sand, and watched.

I got my first clue as to what might be happening when I saw sweeping above me a mighty monster with a huge hunk of meat dangling from his undershot mouth. As he swam, a stream of red blood streamed out behind like the trail of a burning plane.

Ten minutes later, the excitement seemed to be all over. A few dawdling sharks drifted to seaward, and in several minutes, the water was as quiet as when I arrived. Waiting a little bit longer, I made a survey and found that I was all alone, and I started back up toward the surface.

I pushed the camera up out of the water in a lunge over the gunwale and then pulled myself up over the side of the boat. Slipping my glasses back off of my head, I saw my fisherman friend sitting there staring at me as if I were a ghost. His eyes were as big as saucers. His jaw was hanging limp. Like a zombie, he prepared to go back to the shore, not uttering a single sound all the way in.

That evening, I went to a friend whom I had found who could speak both Spanish and English. "What in the hell does *tiburón* mean?" I asked him.

"Shark!" he answered as his eyes dilated.

"What does *marsopa* mean?" I inquired.

"Porpoise," he replied.

The light came. I went back to the schooner and got out my little vocabulary card on which I had copied a selection of words I wanted to remember. There it was:

Tiburón—shark

Marsopa—porpoise

There was nothing wrong with my *Learn to Speak Spanish in Five Minutes*. I had made no mistake in making up the vocabulary sheet. It was just that my memory had been a little wrong.

My memory had been so bad that I had forgotten how bad it was!

The next day, I hiked down to the beach to the corresponding point on the land. The mystery was soon apparent.

The shed on the beach was a slaughterhouse. Every afternoon at four o'clock, their closing time, they dumped into the sea the refuse of the day's work—hundreds of pounds of the residue of the slaughtered cattle—heads, entrails, blood, and bones. It all went down a little chute at the end of the pier.

The word had probably been passed around in the shark schools for years. You could set your watch by it—they were always there at four o'clock, Dominican Standard Time.

The expedition had been a decided success. I hadn't been eaten alive by the sharks. I had had some wonderful tropical diving experiences. I hadn't found any substantial treasure in the old Spanish galleons, but as I headed back for New York, I had in my suitcase thousands of feet of beautiful underwater color film for my forthcoming lecture tours.

36

LIMEY TANKER

The skipper of the *Empire Mica* had been faced with a serious decision. It was night. June 29, 1942. He was drawing thirty-four feet of water. He could not carry that draught into Panama City. It was possible but certainly not advisable to try to bump her through the thirty-two-foot channel into Port St. Joe. He could drop anchor, but he would be like a sitting duck.

Although his decision was much disputed later, he decided to head for open sea.

She was another of the long line of tankers that had been streaming out from the rich oil ports of Texas, laden with the vital fuel of war for the far distant beleaguered island of Britain.

She was carrying approximately six million gallons of one-hundred-octane gasoline, the precious fuel which was feeding the big bombers in their nightly devastating raids on key German industry. The *Empire Mica* was riding so deep that her main deck was constantly awash even in the moderate sea that was running that night.

Suddenly there was a terrific explosion midships. The huge ship shook like a leaf under the impact of the powerful Nazi missile of destruction. Still shaking, with scarcely time to catch her breath, again streaming through the dark water with almost perfect aim came the second torpedo toward the wounded ship. It struck a moment later at almost the same spot, tearing again at the starboard side to make what could only be a mortal wound.

Shortly after noon the next day, almost twelve hours after she had been struck, her bow dropped beneath the waves to come to a quiet rest in 105 feet of water.

Two years and a couple of days later, we headed our bows out on the same course that they had steered that night back in 1942, southwest by south, a half south from the Farewell Buoy at Apalachicola. We had taken an early start, for it would be over a five-hour run from the dock.

Gasoline would make good salvage. It was not that it was so valuable for its weight or for its volume as would be gold, for example, but there was so much of it and it was quite easy to recover.

It did not take advanced mathematics to get an idea of the value of the cargo. If the one-hundred-octane fuel would sell for around half a dollar a gallon and there were six million gallons down there, the potential gross profit was certainly appealing. We would not get it all, of course, since the torpedoes would have damaged one or two of the compartments. However, there were dozens of these separate fuel storage chambers, and we could well afford to miss a couple of them.

The most appealing part of the job would be the ease in removing the cargo. Since gasoline is approximately eight-tenths as heavy as water, it would have a decided tendency to float. With the wreck at a depth of approximately one hundred feet, the hydrostatic pressure of a column of gasoline that we would be pumping out would be eighty feet. Thus, if I could go down there and connect a hose or pipe from the surface to the deck valves on the tanker, we would actually have a geyser that would shoot twenty feet above the surface of the sea.

A geyser of one-hundred-octane gasoline! We wouldn't even need a pump—all we would need is a good valve so that we could shut off our geyser after our surface lighter's tanks were full. It was thus a very appealing business proposition. We could charter, at very nominal rates, any number of tank barges that would carry several hundred thousand gallons. We had our own diving boats and diving equipment. We certainly could find the wreck. We could probably salvage the hoses and piping from the wreck itself. Perhaps this was the pot of gold at the end of the rainbow. Even if it was somewhat smellier than the precious yellow metal, at least it appeared that our chances of getting it were good.

Kosta Buzier, veteran fisherman, looked at the clock as we passed this last buoy. We had ample time to make our rendezvous.

Through the gracious courtesy of Chief Warrant Officer Edmund Kenney and the Army Air Force Field at Apalachicola, a plane would fly out over this location at two o'clock that afternoon where, as closely as possible, we would also be. If they could see the outline of the huge hulk from the air, they would drop a smoke bomb. We would then know exactly where to go.

It took us five hours to get out there. The plane would make the same trip in seven minutes.

To be sure, a couple of minutes before the appointed hour, we heard coming out of the trackless wastes of blue sky and water the faint hum of the high-speed aircraft motors. A minute later, we could discern out of the blue the enlarging form of an airplane heading toward us.

It circled overhead, apparently acknowledging our meeting with a dip of the wings, and then flew out in an irregular pattern over an area that would take us weeks to cover with a single grapnel or dragline. The plane finally headed inshore and disappeared into the blue out of which it had come. We decided that they could not see the wreck. It was no wonder, since the water beneath us was a deep blue-black, being as clear as it normally would be but without radiating the color of a very dark muddy bottom. If we had enjoyed a white sand bottom instead, the dark hill of the wreck would have stood out brilliantly in contrast from the air.

We were far out of sight of land with not as much as a wisp of a cloud to break the empty lonesome world of water and sky. It was a futile feeling—we had no bearing from which to search—every wave looked exactly like the next and we were quite sure that they were not standing still. However, Kosta had lived his life on the water, and to him we were not on a trackless nothing—he had an extra sense not apparent to most landlubbers—his sounding lead.

Overboard it went, Kosta feeling the bottom with the skill with which a great pianist might finger his keyboard. Hauling in the lead, he scrutinized the wad of soft Octagon soap that he had pushed up into the hollow bottom. It was dark sand—seventeen and a half fathoms (105 feet). This placed us on a very definite curve running almost parallel to the Florida coast. It did not tell us just where we were on this curve, but we could, he explained, run back and forth along it, the depth at which she had sunk, and we would run across her. Swinging the lead, we thus would have a single dimensional problem.

We chugged along, Kosta swinging out the lead. "Seventeen and a quarter," he would call, and we would head a trifle offshore until we were at our perfect depth. A little later, he would call, "Eighteen less a quarter"—we were out just a little too far. Behind us, the grapnel was dragging just off the bottom. It would be impossible for us to pass over the wreck without fouling the line.

The wind was dying a bit as the afternoon progressed, and as it did, we witnessed a phenomenon that even a fisherman wouldn't see unless he were looking for it.

Off to starboard was a slick. It was scarcely noticeable, but in a definite ribbon-like area, the water was smoother than the water around it. We chugged over to it and found that we could clearly see, when we studied it, a slight glassiness on the sea. A study of the tide indicated that it was running with the slick and almost dead against our bow, and so we decided to follow the mysterious ribbon to its source.

From up aloft in the crosstrees, the slick was even more discernable, flowing like a river on the sea. As we headed upstream, it was getting thicker and thicker—even the faint smell of gasoline was apparent. Soon little globs began to appear, small patches of changing concentric rings dancing on the subdued waves with every color in the rainbow.

In about fifteen minutes, we seemed to be at the very mouth of the smooth river, with only choppy waters ahead of us on each side. We all had the uncanny feeling as though deep down below our keel lay the mighty hulk of our ship.

Lindsey sang out from the masthead, "Oil coming up." We proceeded a bit farther and there, every few seconds, would break a glob, coming up in a little ball and then spreading out in a dazzling array of colors as it hit the flat surface.

Kosta swung out the lead. He turned and grinned. "Eight fathoms," he called. There was a sudden change in depth from seventeen and a half to nine and a half fathoms, or forty-five feet. It *could* be a submarine cliff, but not an inch of this bottom was strange to Kosta. He pulled in the lead. "Burned paint—rust—it's her!"

We hooked our grapnels into the wreckage and tossed overboard the buoys, thereby establishing our claim to the wreck. Even though it was still forty-eight feet away, I now legally owned one of the biggest tankers afloat—afloat as of two years ago. Kosta's boat was not fitted out for diving, and so we headed back to port to get our own.

Nohl at the helm of the *Hermosa*.

Two days later, we were back on the *Hermosa* with full diving rig. It was apparent that Nature had decided that it was not our destiny to continue to be favored. Snarling little clouds raced furtively by at various cross levels in the high sky above us. There was in the air the indescribable sense of oppression, my psychic barometer telling me that there was something evil in store for us.

However, we had come a long way to get out here. These were the late summer days when we could count on a bit of a storm every day anyway. I certainly was going to take a dive—but with every precaution not to tangle in the wreckage so that I could leave on short notice if necessary.

We had noticed a number of barracuda swimming nearby. As I saw them, I recalled how thousands of times I had explained to question askers on my lecture tours (who presented the number one question, "What about sharks?") that sharks are really not a danger to a diver and that probably most so-called shark accidents were actually caused by barracuda.

A barracuda is as ugly a creature as the wildest-brained artist of horror story cartoons could create. With a tough-looking protruding lower jaw, he is the composite of everything mean-looking in a bulldog and a pickerel or black bass. Like a trout, he is a game fish. He will strike at a flashing piece of cellophane but completely ignore a quiet worm. Many a time, I had played a one-hundred-pound tarpon until we were both almost exhausted, and then, as I made a last effort to gaff him over the side of the boat, had a barracuda strike. Swimming at express-train speed (which I understand has been timed at seventy miles per hour), the barracuda would make nothing but a flash in the water. With such velocity would their barbarous teeth strike the defenseless tarpon that his body would be shorn completely off, leaving only a head on my hook.

I wondered what resemblance there might be, in the barracuda brain, between a tarpon on a hook and this diver on the end of his lifeline.

However, after all of this, I decided to take a chance. I had a brand-new black rubber diving dress, which should attract a minimum of attention. The copper helmet was so corroded from years of use that its greenish black exterior was certainly not flashy. I also had a pair of black cotton gloves, which should cover the only bright spot left—my water-soaked "dishpan hands."

The equipment was all in excellent condition, everything being practically new or overhauled to new condition, from the vessel herself to the last detail of the diving equipment. There would be no excuse on makeshift repairs or old equipment.

Lindsey had just completed construction on our brand-new diving ladder, on which I was now descending for the first time. I had previously noted that he had made a mistake on my recommended dimensions, placing the rungs fourteen inches apart, as in a standard ladder, rather than the specified ten inches for a diving ladder. Divers, with their forty-pound pairs of shoes, find it difficult to take big steps. However, it would get me down into the water and, in addition, six feet under. I could stand on the bottom rung just as my head was going under the surface. Diving ladders are invariably made too short underwater.

I groped for the descending line and, as I gripped it, swung my weight off the ladder to the sturdy manila rope just as my window was submerging.

A streak of terror struck through my body as I looked out at the submarine view.

Thousands of barracuda were poised, each motionless, like tigers waiting to strike. Without another thought, I swung around and grabbed the diving ladder and prepared to head right back where I came from.

As I started up the rungs, I thought of the countless hours we had spent getting this far—long hours in preparation, overhaul, and at sea under the tropical sun; fighting storms; and working all of our daylight hours against the many odds that wartime restrictions can place on such enterprise. Down there, only forty-eight feet beneath me, was the main deck of that giant tanker.

If the barracuda were going to strike, it would probably have been all over by now. I decided, instead of crawling up on deck and quitting, that I would just stand on the bottom of the ladder with my helmet underwater and watch them for a while. I am certain that within my range of vision—which must have been almost forty feet—there were at least two thousand barracuda. They were all full-grown, running about six feet in length. For ten minutes, I just stared and wondered.

If I lost an arm or leg, my recovery of the entire merchant marine of the world would offer little personal compensation. On the other hand, I had selected diving as my business not because of its security but because of my enjoyment in the playing of a game of chance. Every argument I could think of told me not to go down—but down in that mysterious blue-black world was something that was drawing me. And so, to hell with the barracuda. I would progress slowly and cautiously.

With as little movement as possible, I easily slid down my line. The pressure was routinely blowing through my ears, the air compressor was humming sweetly, and all was well.

Down a few fathoms, I squirmed to see what might be below. There were thousands of barracuda, but now—

Looming up beneath me was the giant black corpse of that mighty ship. My heart shot up into my mouth as I viewed her indistinct form.

You can have your palaces and penthouses, your bridge clubs or your bars, your cities and your suburbs—here, in this unexplored world, again, was the thrill that I had all my life sought.

To me, a ship is far more than so much steel, brass, and wood. From the moment she hits the water, as she slides down the ways, to the day she disappears beneath these green waves, she is a living, vibrant being. This ship was even more than that, a faithful servant lugging the fuel of war halfway round the world in our fight for the very existence of civilization. She had died a heroine in these dark waters—and here I was, viewing her mighty dead body, motionless in her blue-black tomb.

Progressing a few fathoms farther, I could now see the details of the wreckage on which I was landing. Torn, mangled iron protruded everywhere—perhaps this was where the torpedo had struck. A few minutes later, I felt my heavy diving boots strike against solid steel plate.

So thrilled had I been with the sight of the wreck that I had momentarily forgotten about the barracuda. They had not forgotten about me, however. As I came to rest, I was very much aware of their presence. One of them, almost like a playful lovesick cat, was brushing against my upper leg. I looked out through the upper window of my helmet and was startled. I was not only in a new underwater world—I was in a world within that underwater world. Above me, almost like an unbroken sky was an apparently solid unbroken hemisphere of barracuda. There were thousands and thousands of those grim creatures forming almost a solid ceiling of fish. Had I the harpoon gun and shot in any direction, it would have been impossible not to hit one.

Looking upward, I could see nothing of them but their light-colored bellies. Looking out of the front and side windows, I could see that sky of fish coming down to meet the horizons of this wreck in all directions. I was really surrounded.

Looking at their vicious faces, I imagined that they had been attracted by the smell of the gasoline. Why a fish or anything else should be attracted by gasoline, I'm sure I don't know, but they looked just evil enough to take a perverse delight in such unsavory stuff. There were more barracuda right now over my head than I imagined lived in the entire Gulf of Mexico. I had often seen them, two or three at a time, on the reefs, but I was at a loss to explain why there should be such a teeming convention on this one wreck.

I started to explore the mammoth ship. I was standing on a pile of jagged, twisted, steel wreckage. In the two years she'd spent down here, she had grown over with marine slime that made the footing extremely slippery and dangerous. A tide current was running over the wreck that added considerable uncertainty to my movements. If I should slip and fall almost anyplace, my diving dress would be torn like a piece of tissue paper.

The best thing to do would ordinarily be to go back to the surface and then drop down on another portion of the wreck. This would be a lot safer than crawling over the mass of jagged wreckage. I apparently must have been at or near where the torpedo had struck. As I looked at the torn, crumpled steel, I was amazed to observe the terrific force that missile must have had.

However, as I thought a moment longer, to go up and come down again would mean going through that teeming sea of barracuda. I would eventually have to go up, but another added round trip through them was just too much to consider. Accordingly, I decided to try to crawl gingerly over the slimy sharp wreckage.

For the entire balance of my dive, about an hour on the bottom, I crawled along very cautiously and slowly. I felt like a mountain climber working his way up a sheer icy slope, moving inch by inch and painstakingly testing each new purchase. One careless slip, and I might rip my dress into ribbons.

That must have been some torpedo, I thought, as at the end of an hour I had not emerged from the shattered town plating. I had been crawling over not only the shattered steel but also a web of tangled wire rope, deck machinery, huge valves, and stanchions—almost everything of which a huge ship is made. The destruction that I had passed over seemed far more than could be caused by any torpedo that man had ever invented.

I could feel a little wave of action. At this depth, it must mean that quite a sea had picked up. It was probably a good thing that I was heading back, since there could be no more certain death than to have the schooner drag anchor and tow me through this wreckage. It would be no better than passing through a mammoth meat grinder.

As I approached the surface, I began to sense the magnitude of the storm. The descending line was jerking violently up and down, the bowline around my waist jerking me violently at the bottom of each stroke. A squall must have come up, which at this time of year could grow to nasty proportions.

However, of particular worry to me at the moment were the barracuda. Looking down and up, I realized I was now the center of an almost perfect sphere of these vicious-looking fish. I was dangling like a piece of bait on a string and, as I hung there in midwater, was completely exposed in all directions. Those ugly monsters were studying me closely no matter which way I looked. There wasn't much I could do about it. I had to go up and take my chances. I certainly couldn't stay down in the wreckage with the storm coming up, even if I had had more time.

Suddenly, there was a terrific jerk, as if a giant whale had struck my lifeline above me. I was swept up toward the surface and, in a moment, found myself being towed at breakneck speed just a few feet under the surface. My first thought was that one of the barracuda had struck my lines for some reason. However, that didn't make sense—and as I thought it over, I realized that the *Hermosa* had lost her anchorage and was racing free at the mercy of the wind. As I later found, she had dragged her biggest, heaviest anchor into the wreckage and torn the line on the sharp plating—we were adrift.

I had a good start on my decompression but could see that I would have to get along without the rest. To offset this worry, I noticed that I was now being towed out of the barracuda colony into clear water. Two or three minutes later, there was not a fish in sight. I had not been eaten.

Apparently, there must have been trouble topsides because no one seemed to be paying any attention to me. The water was just racing by, indicating that we were being blown at a terrific speed.

In a moment, I heard the whirling of the propeller. I was hoping that they had taken precautions not to throw her into gear until my lines were clear of the whirling blades, since this was no time to get my hose fouled there, both for my continued health and for the much-needed use of the power in keeping the vessel's eyes into the gale. Apparently, she was swinging into the wind. I felt a momentary slacking off of my lines and a little tendency to sink to my weight. Rolling over, I could see the underbody of the *Hermosa*. Her wheel was revving up, and the slipstream was deflecting off the rudder as the latter was thrown hard over the port before my eyes.

In a few minutes, I felt the boys pulling me into the hull. I squirmed and turned to reach for the ladder to help myself out of the water and was quite surprised—there was no ladder!

As I later learned, it had been torn from its fastening, the half-inch bolts having been pulled out as if they had been made of toothpaste.

The next few minutes were the very bad ones that many a diver has experienced when he has tried to board a vessel without a ladder. The boys were all hauling away on my lines, there being no shortage of willing muscles available. However, my weight plus that of the two-hundred-pound diving suit all being supported, over the side of the madly rolling boat, by the noose of rope under my armpits was not exactly the ultimate in comfort. I grabbed the gunwale and contributed as much as I could in supporting myself in the necessarily awkward positions. It was murder, but they finally made it by reaching down and unfastening my lead belt and then lifting this on deck as a separate item.

In what seemed like ages later, I was lying flat on my chest and belly with legs still bending over the side. My tender was taking off my helmet.

Suddenly, I realized that I was entering a very different world than that which I had left behind a couple of hours ago. The instant that the helmet came off, I thought that a battery of machine guns had suddenly been turned upon me. Although it took me a few seconds to realize what it was, I soon discovered that it was not lead but drops of water traveling at the speed of a bullet that were hitting me.

I had heard of and seen pictures of pieces of straw that had been driven through fenceposts in a tornado, pieces of paper and chips of wood that had done the damage of a rifle bullet, but never had I experienced or fully realized the power that can be vented by a hurricane-force wind.

The boys helping me had their heads securely wrapped in heavy leather jackets. They could not expose their faces to the peppering blast of rain-drop projectiles without danger of serious damage to their eyes and faces. As they removed my helmet and a few snipping shots hit me on the neck, they tried to give my exposed head temporary cover with a thick wool coat, which someone had brought up from below. In the vicious wind, it would not stay fast without elaborate lashing operations, and the boys could stand it no longer.

"Put the helmet back on," I asked, sensing that this would best stand the barrage of water bullets. With the front window open, they did this, and I just fell forward on deck and waited. The boys disappeared.

One of the boys had done some fast thinking. The big heavy diving ladder, made of two-by-sixes, had been broken off its mounts in the storm and, unless quickly lashed, down it would soon be swept overboard and lost. No person could stay on deck and man the wheel to keep our bows into the mammoth seas. Combining the responsibilities, they took our heaviest manila anchor cable and made it fast to the center of the middle rung of the diving ladder and then tossed it overboard. They let out about two hundred feet of line, ran it through a fairlead, and then made fast to the bitts on the bow. We thus had a very effective, if quickly improvised, sea anchor.

This had been a fortunate decision. Through the side window of the helmet, I could see the effectiveness of that rig in keeping our bows into the wind and seas. At the fury of the gale, our schooner would not have lived for five minutes should we have exposed our side to those vicious waves.

If I had never prayed in my life and there was one time that I would have wanted to, this was it. The anchor bitts were solidly mounted to the frame of the vessel such that the entire bow would almost have to pull off before they would give. The manila line was an inch and a half in diameter and brand spanking new, so it should be good for its maximum strength of many, many tons. The diving ladder was extremely heavily built, the massive rungs being made for the heavy weight of a fully equipped diver hanging over the side of a rolling boat. It was heavily bolted together without a nail in it, and the danger of its pulling apart was small. We should ride out the storm.

We did. What a false economy it can be, as I had learned many years ago, to go to sea with flimsy, improperly designed equipment! Other than the loss of our anchors and the wrenching off of the diving ladder mount, we came through no worse for the storm. The fury of the hurricane abated after about ten minutes; thereafter, it was just a violent gale, which, after an hour, dropped off to just a bad storm.

However, the boys were all a little frightened, including me, and we were anxious to get our feet on solid ground. A storm like that is always hard on everyone. The galley was a mess. Though we had no casualties of equipment or even dishes, food had been hurled out of containers and spewed all over the room. Personal belongings needed drying out. The Gulf would be milky for several days anyway, as a result of the churning action of the water. We headed our bows into the Port of Apalachicola.

Three days later, we were over the *Empire Mica* again with fresh energy but with a big question mark in our minds about what had happened to the wreck. The Navy Department had been very cooperative in giving us highly restricted information about her. We had been assured that no salvage or demolition operations of any sort had been conducted on her.

I dropped down on the giant ship again in another place. Again, I was only on a mass of torn, jagged steel.

Again and again I dove, landing on various parts of the wreck. In every case, she was badly torn up. Unless she had been struck by a thousand torpedoes, and we knew it had been only two, something had happened to her since she had sunk.

It was obvious that the great gasoline compartments of this huge tanker had all been ruptured. The valuable high-octane fuel had leaked out into the sea.

I prowled over the wreckage a little and picked up a few things which had resale value and which would just about defray our expenses. The pot of gold at the end of the rainbow had again eluded us.

We found out the story later. Unbeknown to the people to whom we had talked, the *Empire Mica* had been dynamited. Although the wreck itself had been lying in deep enough water so that she could be of no conceivable menace to the deepest draft ship afloat, unfortunately, she had landed on the bottom in a perfectly upright position and her tall masts had protruded vertically upward so as to almost break the surface. From this standpoint, she was a possible menace to navigation, since those massive steel spars could do serious damage to the underhull and running gear of almost any passing ship. Although it seemed to me that they could have been pulled out by the power of a tug or blown out with a modest charge of dynamite, the decision had been to blast the entire wreck to smithereens. Because of the veil of wartime secrecy, these things were not as widely known as they might have been in years of peace, and we were unintentionally misinformed.

Now that we had learned the channels through which this information was handled, we investigated some of the other wrecks that we planned to salvage. There was evidence that demolition operations had been conducted on our other selected prizes, too.

We were having trouble on every turn. I had had a terrific fight to get permits to operate my commercial boats in what was then considered combat waters. I had to get permission for every move that I made, and I never knew how long it would last. I had to get special permits for gasoline to run my engine and compressors. These coupons had been metered out to me with a lot of talking on my part, but I never knew where the next gallon would come from. We had just so much diving equipment left, which was made of extremely critical rubber, copper, and brass. Should we have a few falls down on that jagged wreckage, we would be out of equipment and could only get more by a special dispensation from God or Mr. Roosevelt. Should we have trouble with our engines, getting spare parts would take almost an act of Congress. Worst of all, we didn't have a single able-bodied man in our crew as diver or seaman. We had made out thus far with the best selection of 4-Fs that we could find, but this was not the right way to tackle the big job ahead.

We were just swimming against the current. The wrecks would have to wait.

37

Underwater Honeymoon

Time has a way of soothing all things, and so my irritation about the snarl of government restrictions quieted down. Soon, along the Gulf Coast I was again hearing reports of the merciless torpedoing of our merchant ships. That gleam was again conspicuously showing in my own eye.

The German submarines were moving westward, catching our mighty ships only hours away from their Texas loading ports. I now had definite locations of nineteen merchant vessels, all over ten thousand tons, lying off the coast of Louisiana. That was just too much—I had had my trouble with the *Empire Mica*, but I believed this expedition would be different.

It wasn't.

The proverbial boy who burns his finger on the hot stove is supposed to learn not to touch it again. Apparently, I didn't learn.

This time, however, I was more cautious. I went down to the logical base port for this area of the Gulf—Morgan City, Louisiana. I talked to fisherman after fisherman. Certainly, the wrecks were there! If we couldn't find them from the special restricted charts that the Navy Department had given me, the fishermen could. They knew the dramatic stories of the sinkings and had helped with the rescue of the few handfuls of survivors. They knew the locations well enough to stay a long way away from them; catching one of their nets on such a wreck spelled nothing less than total loss for that very expensive piece of equipment.

They admitted that they steered far clear of them, but on curious little sallies, with their nets not overboard, they had prowled in the vicinities of the slumbering giants and had seen the oil slicks coming up. A few of the huge hulks, caught in shallower water and now lying upright, still had their mastheads dramatically protruding above the surface.

I asked the boys to bring over the sixty-foot *Hermosa* from her base port at St. Marks, Florida. She was completely fitted out for diving and could accommodate a dozen men nicely. I started to set up our offices and ship in the small Louisiana port that apparently would be our home for some time.

I was fascinated by the little town. It was not much more than a high spot in the vast coastal swamp area bordering the Gulf Coast. Though the town sat on the broad muddy Atchafalaya River, I was surprised to discover that, although it was known as a Gulf port, it was about a thirty-five-mile run down that muddy stream before a boat could get into open Gulf.

As I studied the charts, I began to wonder where Louisiana actually stopped and the Gulf of Mexico started. The vast swamp area to the south of us was neither land nor sea. It could be penetrated only by boat; the native pirogue was able to navigate almost anyplace through the musty swamps where the water was over an inch or two deep. The maze of what seemed to be islands was not solid ground at all, but merely illusory combinations of floating vegetation and oozy mud on which only a bird or insect could walk.

As the traveler ventures farther south in his pirogue, the apparent water spaces widen more and more, and the islands grow smaller and smaller. Finally, a trace of motion, suggesting the waves and swell of the Gulf, is felt in the water as the cypress trees and marsh grasses thin out. The pattern gradually changes from water patches between mucky floating islands to smaller and smaller islands in larger and larger patches of clearer and clearer water. Finally, there comes a place where only occasional wisps of islands remain, through which can be seen the great unbroken blue expanse of open Gulf to the south. Where exactly the continental coast starts and the Gulf stops is difficult to determine.

Morgan City, a high spot of ground in this great trackless swamp, boasted proudly that it was the largest shrimp port in the world. In the warm Gulf waters off the Louisiana coast were the beds of the famous jumbo shrimp, the largest and finest known to connoisseurs of the seafood delicacy. Hundreds of fine trawlers made their base port there, traversing the long muddy river through the swamps for scores of miles, as they went out and came back from their one- to four-week shrimping voyages.

As I waited for the *Hermosa* to make the thousand-mile trip from Florida, I realized that I would have to have a place to live. This was 1945, the last year of the war, and the shrimp business was booming as part of the increased wartime demand for food to supply the Allied nations. Even after the arrival of my salvage boat, I thought that I would prefer to have my living quarters and expedition office on dry land rather than in the smelly crowded quarters of the vessel. Morgan City was now also emerging as the main base port for the drilling and exploring operations on the offshore oil wells. In brief, the town was greatly overcrowded. There just were no quarters available to rent.

I couldn't sleep on the dock in that mosquito-infested country, so I decided that I would have to make it my business to find a place to live.

Persistence finally brought results. Quite accidentally, I found a very attractive upper duplex, which was available immediately, but there was one hitch. The owner positively would not rent it to an unmarried man, which she assumed I was; all the unmarried men she had ever known were "so sloppy," and although she said she thought I was very nice, she was afraid that her property would not get the right care. I could see that the rooms had just been completely and handsomely redecorated and were in beautiful condition.

"Oh, but I *am* married," I said without a second's hesitation. "My wife is up in Wisconsin," I explained. "We thought it would be best for me to find a place to live before she came down, since we guessed that a home might be difficult to find. She'll be down as soon as I find a place to live."

Apparently, that satisfied the lady, and I took the duplex, paid my rent in advance, and moved in.

This created an obvious problem. The owner lived downstairs; in not many days, she would be asking about my wife. I didn't have any. I wasn't even engaged.

I was thirty-five years old. My friends had all written me off as a hopeless bachelor. How could any woman put up with the type of life that I had always led? I was either out on expeditions, on diving jobs, or, in the winter recesses from these, on coast-to-coast lecture tours. A wife would want a home.

When I had been in Milwaukee, I had devoted a lot of attention to a young lady by the name of Eleanor Hecker. I had known her in high school eighteen years earlier (she had been a few years behind me), but I hadn't seen her again until just recently. I had thought all of my old friends had long since married, settled down, and gone to seed. She hadn't.

We had been getting along famously and had had some wonderful times together. If I *ever* wanted to marry, she was the woman for me. Of this I was very sure.

But I wasn't sure about that home business. I had been free as the wind for all of these years, and I was in the habit of being single. At thirty-five, this was a well-entrenched habit. I had thought seriously of getting married before leaving on this expedition, but while I was considering it, all of a sudden I had to leave.

However, now I saw the entire matter in a new light. Here was a beautiful place to live, in this very interesting little town. Even though I would probably be out on the boats at frequent intervals, I would also spend much time ashore on the business end of running an expedition. Furthermore, here I was in a lonely little swamp town made up of fishermen and swamp trappers. It looked very much as if I might be here a long time—the floor of the Gulf was cluttered with torpedoed ships. And I could certainly afford to get married—business had been very good indeed.

That was the answer. I would get married. All of a sudden, it came to me like a flash that night. I never doubted my decision for a moment at any time after that.

There was only one possible difficulty. Would Eleanor be interested? I had never proposed. We had never discussed the subject. There had never been even a hint of any such plans. But I did have the feeling that something had just sort of clicked when we were together.

That evening, I got out my portable typewriter and wrote her a letter. I could propose for the price of an airmail stamp. I worked on the letter until late that night. As soon as it was finished, I grabbed my hat to go out and mail it, but then decided that I should wait until morning to drop it in the box. It wouldn't be picked up until then anyway. If I were still sold when I awakened, I would mail it and then never change my mind.

A week later, I met her plane in New Orleans. We picked out a pretty vine-covered church on St. Charles Avenue in which to speak our vows.

Eleanor had always expressed much interest in my diving. I knew that she liked travel, at least in moderate doses. She would certainly fit into the picture.

That very day, before the man said, "I pronounce you man and wife," we talked about the future, the type of life that I had led, and the life we would lead together. We could still change our minds.

But we were in agreement. I told her honestly my decision that I wanted to get away a little from daily diving and expedition life. I would like to get into some more basic phase of the business. I still wanted to go out on an expedition at least once a year and would like to take a limited number of lectures each winter, but somewhere I wanted to set up a home and see to it that I got there once in a while.

She liked this plan, too. She could then go with me on some of the expeditions and also on some of the lecture trips. On the diving expeditions, she would like to learn to dive, and on each of these trips, she could do some diving.

We were in perfect agreement. We got married.

This was no time for a honeymoon trip, however. The *Hermosa* was now in at Morgan City, and we needed to get started on our investigation dives on the wrecks. I would take my new bride with me—she certainly would enjoy the interesting trip down the river through the Louisiana swamp country, as well as the subsequent open-sea sailing.

We made the run down the river and were just about at the mouth, ready to push our bow out in open Gulf. It was now late in the day, and there was not time to accomplish much out there today. The Gulf was kicking up a little; it seemed advisable to wait until morning to go ahead. We could drop our hook in the river and thus avoid what might be an unpleasant and unnecessary night spent riding at anchor in a heavy sea.

The water was only two fathoms (twelve feet) deep here. The boys were checking over the diving gear, taking advantage of the river calm to complete last-minute operations.

It occurred to me then that here was the chance to let Eleanor take her first dive. Conditions were ideal; we certainly would not want to let her try it for the first time in deep water on a wreck lying in open sea. Eleanor agreed that this would be a good place to start.

We very carefully helped her get into the suit, making sure that everything was adjusted for maximum comfort and ease of handling. In our courting days, she had asked me a lot of questions about diving. She had even looked up a good bit of information on it at the library, in true Dale Carnegie style, so that she could discuss it intelligently. This was all very impressive and convinced me, as I thought about it later, that Mr. Carnegie was certainly right. I quickly reviewed with her her book-learned information and found that she remembered the basic principles involved. She checked on the positioning of the valves so that she could reach for them blindly in what we knew was going to be dark, muddy water.

The boys were gracious in helping her, holding the heavy breastplate in their hands so that she would be spared this load as long as possible. As soon as she got underwater, the buoyancy of the water would automatically relieve her of the weight.

I could see that she was very much concerned about it, but apparently, she was planning to go ahead with the dive.

Since an overexcited beginner will always require more air than will a calm and collected veteran, I decided to cut in the second engine compressor, either one of which could take care of an experienced diver in 120 feet of water. Thus, she would get her full output of the two big units. There was a blast of air roaring into the helmet that rivaled a small hurricane.

She said "OK," and the boys moved fast to get her overboard with a minimum of discomfort. As they slipped the helmet over her head, she experienced the feeling (which I had forgotten to warn her about) known to every diver—the deafening roar of air, enough to wake up a dead man and then scare him half to death. The boys now needed all their available hands to fit, turn, and lock the helmet into position, and they momentarily let go of the parts of the suit that they had been supporting, thus dumping the full weight of the hundred-pound lead belt on her shoulders. The twenty-eight-pound shoes on her feet were in somewhat startling contrast, too, to the dainty little pumps in which she had been married a few days before.

I cleaned the window on the helmet, which sealed the space within the diving suit. The dual compressors, roaring their output into this tiny interior, certainly would frighten anyone.

Dragged downward by the two hundred pounds of weight, down the ladder she went into the muddy water of the Atchafalaya River and right on down to the soft, oozy quickmud bottom. We had warned her that visibility would not be very good, and she probably had not expected to see as beautiful a submarine landscape as she had observed in some of my tropical underwater films. But she had not expected to be in total darkness before her helmet was four feet under the surface.

The quickmud, total darkness, deafening roar of air, two hundred pounds of diving suit, and blinding pain in her ears and head (from her unadjusted eardrum)—it was just too much. She had a dim recollection that she could do something about that terrible blast of air with a certain valve that she had read about—but that was in the light of day. In the inky water, she fumbled frantically and found the valve. It was the right valve. But she turned it the wrong way.

The additional air rushed into her diving dress and distended it out, giving her a tremendous buoyancy. Up she went like a skyrocket, almost jumping out of the water as she hit the surface. She then floated face down, spread-eagled like a big rubber beach toy, and we towed her over to the ladder.

Eleanor had been going to dive with me on every expedition. These expeditions would occur at least once a year. Therefore, it would be an annual dive.

This was definitely Eleanor's first and last annual dive!

The rest of our time in the Gulf was an exact duplication of our experience on the *Empire Mica*. The wrecks had been dynamited, in spite of the assurances we'd received to the contrary. The terrific blast of the demolition charges had split open the great fuel compartments of the tankers, and their valuable cargoes had leaked out into the water. The cargo of the freighters was still undoubtedly down there with the jagged twisted hulls of the exploded ships. But now, at this point in my happy new life, I would not venture down into that tangled torn mass of razor-sharp steel plate in those blind waters, clogged daily with thousands of new tons of Mississippi mud—not for all the gold in the world. It would be certain suicide.

Eleanor and I headed back up north. I was going back to my routine and free-from-danger diving business.

38

550 FEET

Since the minute in 1937 when my lead shoes touched the 420-foot bottom in the middle of Lake Michigan, Jack Browne had nurtured a deep desire to beat my world record. He had actually told me that, but from the reports I heard from everyone else, I realized that this was no passing fancy.

On April 27, 1945, the great day arrived, following months of intensive preparation. Jack was going to attempt to reach a simulated depth of 550 feet in the DESCO experimental tank in the basement of the company's plant in Milwaukee.

Since I had dropped out of DESCO in 1940, a great world war had come and gone, and the hibernating little company that I had given up for dead had come to life with nourishment from Uncle Sam. I had set up my expedition bases in Florida, but Jack Browne was still in Milwaukee and had stayed with the company.

This tank, I begrudgingly had to admit, could produce the same pressure conditions and could have the same physiological effects as open sea. The tank was approximately six feet in diameter as it stood on its lower dished end and was approximately twelve feet high. It was filled with water to within two feet of the top, making the depth about ten feet.

However, the top was fitted with a heavy hatch. With this closed, air could be admitted into the top of the tank under high pressure. The air would act on top of the water so that the combined pressures could create almost any desired diving depth.

Jack was wearing the DESCO Lightweight Suit, a surface air supply type of apparatus that had no helmet. The rubberized dress covered the entire

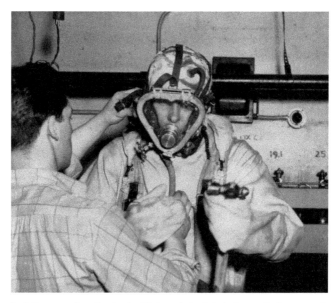

Jack Browne, dressing in the Lightweight Suit in front of a battery of helium tanks.

body including the head. At the diver's face was cemented a mask, which could give him fine vision. Because there was no huge helmet with all of the air space inside, the entire outfit displaced very little water and thus required only a fraction of the lead in the belt and shoes.

The Navy Department was extremely interested in this dive, since this suit was now an official navy specification. They brought in dozens of government tanks containing high-pressure helium. This was to be fed to the suit through a pipe in the wall of the tank, which, in turn, connected to a short piece of flexible diving hose leading to the suit.

Jack accomplished a feat which unquestionably was one of the all-time great landmarks in the history of diving. He reached a simulated depth of 550 feet and submitted his body to the corresponding pressure, approximately 240 pounds per square inch, for thirty seconds. His 550-foot dive topped my 420-foot descent by 130 feet.

Jack's dive stirred up quite a bit of controversy. My followers maintained that I still held the official world's record and that Jack's feat could not be accepted until he duplicated it in open sea. He had everything under artificial and ideal conditions, they maintained. He had comfortably warm water, electric lights, big windows all around the tank, on the outside of

Jack entering tank through top hatch, showing pressure valves and gauges.

which, only a few feet away, were all the diving experts of the US Navy and the distinguished authority on high-pressure physiology, Dr. Edgar End. After a short decompression, Jack had been removed from the testing tank, placed on a stretcher, and slid into a portable DESCO decompression chamber. This was pressurized and then loaded onto a truck. In this way, Jack was carried into the portable one-ton chamber that we had used on our helium tests years before and that we had also used on the disastrous twenty-seven-hour dive. Inside of this huge chamber, he could enjoy all of the luxuries of his home during the balance of the nine-hour decompression.

These followers maintained that I had to do my dive under actual diving conditions—ice water, total darkness, working from a rolling, pitching vessel, and in another world, all by myself. They pointed out that on all records, exacting conditions had to be met. On all land, water, and air speed records, for example, the course had to be transversed twice, each time in opposite directions.

This was all very gratifying, but I knew that Jack had really done it. Certainly, the water had been colder on my dive, but I hadn't been cold inside of my lambswool fleece underdress, so how could that have affected the end results? Jack had been so close to the experts that he could almost touch them—but that was a big "almost." On either side of the one-inch-thick steel tank, it was possible for them to be as close as one inch apart, but in pressure they were 240 pounds per square inch apart, and there was no possible way that their presence could have offered any more than psychological assurance to Jack—though it was just one inch, for all practical purposes it was as if he were in another world. The light might have dispelled a little of the terror, but to an old diving hand like Jack, total darkness certainly would have been a routine part of the business.

Jack really did it. The basic question—what would happen to the human body under such a tremendous pressure when the diver was breathing helium?—had certainly been answered. The body reaction would be exactly the same in the tank as in open sea. It was not only possible but practical to dive in such depths.

Frank Crilley had held the record for twenty-two years, from 1915 to 1937. I had it for nine years, from 1937 to 1946. It was now Jack Browne's turn.

What can you do with it anyway?

Book VII

The SS *Tarpon*

39

THE SS *TARPON*

"To hell with the thousands of dollars in percentage—I want fifty dollars in cash!" Thus spoke the begrizzled fisherman, his jaw set firmly and his eyes staring steadily at me.

Everyone along the Gulf Coast knew that the steamship *Tarpon* was carrying a substantial shipment of American money in her safe. There was only one man in the world who knew exactly where she lay—this man standing in front of us. We were prepared to settle for a substantial part of her proposed profits for him to take us out there, but he wasn't interested—he just wanted fifty dollars. We couldn't believe it, but if that was the way he wanted it, it very certainly was satisfactory to us. The payment, it was agreed, would be made following the completion of our first identifying dive to the wreck.

The SS *Tarpon* had, for a number of years, been operated by the Apalachicola Steamship Company in the Gulf coastal service under the able command of Captain Willis Green Barrow. Barrow was widely renowned for his grim determination and his fine, flawless record for carrying his ship through on schedule, come the proverbial hell or high water. Against the advice of his more conservative colleagues, that man would not be stopped by anything, plunging his vessel's head into whatever might come up to make port and make port on time.

It was on the first day of September 1937 that the *Tarpon* was steaming on her regular course out of New Orleans headed for Panama City, Florida. This was in the heart of the hurricane season—the barometer was dropping rapidly, and the seas were building up to mountainous proportions. Barrow was off watch and the ship was in the hands of L. E. Danford, its more conservative first mate.

From the engine room were coming up reports: "We're taking in water, sir." Her first officer was reluctant to disturb Barrow for what he knew would be a refusal to change her course.

Finally, the reports from the engine room indicated that the water was coming in at just about the rate that all pumps at capacity could handle. The report was relayed to Barrow but was met with only a curse to "Stay on your course."

Seams were opening more and more. The latest reports from the engine room, that the leakage was far overtaking the pumps, were needlessly spoke, for the first officer could feel the sluggishness of his helm with the lurching of the added countless tons of weight in the bilges. As he felt her hull settling deeper and deeper into the battering seas, the man finally decided to take matters into his own hands. By changing her course ninety degrees, he could run with the waves immediately and save her from the terrific abuse she was now taking, and, at the same time, head her for the shore which he honestly knew was the only thing left. She could not live over a few more hours under the pounding, and it would be far better for her to nestle on the soft white sand of the beach with comparatively little loss than to risk almost certain sinking in these deep offshore waters. He threw her hard over. The *Tarpon* sluggishly answered to her helm, swinging out of the troughs of the angry seas.

Barrow, belowdecks, immediately was conscious of the change in course by the feel of the deck beneath his feet. He quickly dressed and a few minutes later came storming up on the bridge and demanded to know on what authority the decision had been made. The first officer reviewed the reports from the engine room, the status of the barometer, and the obvious feel of the tremendous weight in water already in the bilge.

Captain Barrow snatched the wheel out of his hand and swung her back on her course. Issuing the oft-repeated statement that had made him famous: "God runs the weather. I run the *Tarpon*!"

Five minutes later, her battered hull came to rest in a blue-green world of peace, 130 feet beneath the angry waves.

Captain Barrow and eighteen of his crew went down with her.

The other seventeen men clung to bits of wreckage and wondered whether they would ever be saved. They were probably twenty-five miles off the Florida coast, but in the storm the visibility was not over a few hundred yards. The chance of their being found was almost negligible. Not even a radio message had been sent out giving notification that they were as much as in trouble.

One of the boys—the second assistant oiler, Adley Baker—knew that theirs was only a fighting chance, and he was the kind of a fellow who would fight until the last breath had left his body. Guessing at the approximate direction of the mainland from the set of the wind, he gave up his precious hunk of wreckage and started swimming.

He swam all the rest of that day. Night came. Desperately he kept on, still estimating his shoreward course from the wind. On and on, he grimly paddled through the long dark hours. Morning finally came. There was nothing he could see but water. On he swam.

Twenty-seven hours later, his bare numb feet felt soft sand beneath them. With scarcely more strength than a piece of flotsam similarly washed up on a beach, he limped out of the water onto a broad white sand beach. Crossing the beach, he found the highway, closely skirting the coast, and hitchhiked a ride into Panama City.

Arriving at the steamship office he reported. "I'm Adley Baker, sir, second assistant oiler, steamship *Tarpon*, sir. The *Tarpon* sank yesterday."

A number of fishing boats were dispatched and picked up a total of sixteen additional survivors who had spent almost thirty-four hours in the water.

Quite a few years had passed since that day, but we well knew that the currency in her safe would not be damaged by immersion in salt water. Our fifty-dollar fisherman was loafing on the deck of our salvage vessel as we ploughed the sea with very little to say.

That man, as have many of these real souls of the sea, had some extrasensory instinct that ordinary mortals just don't possess. He was not interested in looking at the compass (which we were watching intently so that we could reproduce this course if necessary). He didn't seem to be looking at anything—just once in a while, he would mumble, "Point her a little" (into the wind) or "let her off a little." Behind us was now just an almost indistinguishable wisp of land, separating sea and sky, so ephemeral that it was with no certainty that we could be sure that we could see anything but water.

"Slow her down," he cried as he noncommittally glanced about at sky, water, and horizon. What it was that prompted his decision was totally beyond our realm of perception, but until we could prove him wrong, we implicitly followed his apparently heaven-inspired directions.

"Hold her! Hold her! Hold her!" he suddenly rang out, and Cherry threw out the clutch, disengaging our throbbing diesel from the propeller. Another of the boys tossed a buoy overboard so that we could have some more basis for reference. "We're over the wreck!" he cried after making another mysterious check of his unearthly bearings.

It all seemed like a big joke to me, but until we definitely decided to dismiss his almost fantastic offer, we had better stick with him all of the way. We had come a long distance on the basis of his story, and it would be foolish to quit before we had even looked.

We brought out the lead and started circling the locale for a sample sounding. We had drifted off a hundred feet from our buoy and now could conveniently circle it with a series of contour soundings.

"Twenty-two fathoms [132 feet]," came the report, followed a few seconds later by "bottom—sand." Without a moment's loss, overboard again went the lead, and a minute later came a second similar call, "By the deep, twenty-two!"

On they came—"Twenty-two," "Twenty-two," "Twenty-two," Twenty-two," and then like a bombshell, "Eighteen!"

There was a sudden change in depth of four fathoms—it might be a submerged cliff, but much more probably, it was our sunken ship. A moment later came the confirmation of our surmise: "Rust on the sounding line. A few flakes of paint stuck to the lead."

We all looked at that fisherman with a deep sense of respect—almost reverence. There was still the question ahead of whether it was the right wreck, but even if not, we were totally baffled as to how he had led us to this very spot. We searched the shore for possible bearing but could see almost nothing.

Our vessel, fitted for sponge, was equipped with the regulation Greek type of diving equipment, the air being pumped with a compressor running off the main engine on a V-belt drive. Everything seemed to be in good order, and so I slipped into the diving gear and jumped overboard from the bow.

We had had a tough trip and really deserved a day's rest before starting out to look for the wreck. We had sailed all night long and were all well worn out from nothing more than lack of sleep. I had caught only a few nods here and there, between tricks at the wheel and the solving of our numerous navigational problems. I had, through the "wee hours," been a bit chilly, at the mercy of the brisk breeze blowing over the black expanses of the offshore Gulf. I now was noticing the effects of this exposure as I started down in what I had hoped would be a rapid descent.

My head seemed clogged up, and it was apparent that blowing my ears would be necessary all of the way to the bottom. I had tried my trick yawn and found that I could, with violent contortions of my jaw, ease the pain. However slowly it was, on I continued down, hand over hand, on the "scedalia," the Greek designation of the descending line. Suddenly, my ears seemed to clear as if by magic, and I started going down like a plummet. As the water darkened before my window, indicating the increasing depth, my heart seemed to jump a little with the thought that in less than a minute, I would be on the rich wreck lost in this blue-black tomb. However, that happening was a bit premature, for suddenly again the blinding pain struck through my head as my tubes stuck again. I grabbed the scedalia to arrest further any falling and then started the usual antics of my jaw to blow open my ears.

This time they didn't seem to work. As hard as I tried, that cringing pain still persisted. I just hung there with it pressing on my eardrums as hard as I could stand it, and waited. I made alternative attempts to remove it by swallowing, yawning, and blowing my nose into the helmet interior space—but all to no avail.

As I dangled there, I paused to observe my surroundings, which were a rather colorless sight. Probably now, about forty feet beneath the surface, I was in that comparatively rarely inhabited world—midwater. There was not a fish nor anything else in sight except the vast expanse of the turquoise blue water, framed by the round circle of my window. Looking upward through the top window, I could see the shadow of the *Arna* silhouetted against the bright surface light. Looking down was only a dark nothing, an entrancing tomb of mystery that seemed to be luring me on into it. There was no question of sufficient lure, however; it was now a problem of blowing those ears so I could continue downward.

I found that the deep blue of that water, rather than being just an empty void, held rather the same attraction as might the color of a clear sky or of a perfect unblemished turquoise. Its undeterminable depth fading off into a mysterious nothing held a strange fascination. I decided to twist my body around and see if there were any forms of marine life to be seen in other directions.

Paddling frantically with my hands, I imagined myself turning, although I could find absolutely no mark in the deep blue wall by which to judge my relative movement. And then, suddenly looming up not two feet from my window, was a giant tentacle.

I practically jumped out of the diving suit with fright. It was a slimy-looking thing, extending almost vertically before me. It seemed to be at least a foot in diameter and apparently must be a leg of some mammoth monster. It was so close that I could get no perspective on it whatsoever, the slight tilting of my helmet that was possible only revealing another foot or so of its length.

My direction of escape was rather limited. I could not go down because of my ears. Except for a little wiggle, it was not possible to move in any direction horizontally. I could still go up, however, which was exactly what I started to do.

Squarely facing the weird-looking object, I climbed upward by it and soon realized that it was perfectly straight. The light was just barely beginning to dawn on my pressure-crazed brain; suddenly the answer was very clearly revealed before my eyes. There, fastened to the slimy "tentacle," was a huge double-sheaved block (seagoing pulley) through one part of which was still reeved a heavily moss-covered manila rope. This was the mainmast of the SS *Tarpon*. Moving a little farther up, I saw the mast rings to which were fastened the upper ends of her side stays. These led down at their respective angles to the gunwales of the huge ship below.

Moving up another couple of feet, I saw something that sent a tingle all through my body. Mounted at the head of the mast was a black wooden sphere about the size of a man's head, supported on a neck the size of a lead pencil. This was simply an ornament—but I clearly remembered noting that same curious ball at the masthead in the pictures we had of the steamship. It must be our ship!

As I visualized myself like a little flea hovering high in space near the top of the mast of this ship, I realized that I could now draw a few conclusions. For the stick to be standing, the wreck could not be in too bad a condition. From the vertical angle of the mast, it was gratifying to conclude that she must be lying almost upright on an even pitch and even keel.

Thus far, I had not been able to actually touch the formidable structure before me, but now, with a violent swimming motion of my hands, I was able to grasp the huge spar. Another frantic lunge and I had my hand on the side stay. I didn't know whether this was the port or starboard side, but here certainly was an excellent guiding line to a definite part of the ship.

The clinker in my ears was now apparently dislodged, for I was able to drop down as fast as I dared slide along the somewhat frazzled steel stay. Moving another thirty feet, I stopped to find out if I could see where I was going. Bending violently over with the help of the taut stay line, I could now barely discern the dim outline of the deck beneath me. There, as if shrouded in fog, lay the corpse of this once-proud ship, covered over with green slime and littered with wreckage everywhere.

Now within ten feet of the deck, I let go of the stay and dropped free to land only a few steps from the cargo hatch. Instead of a firm deck breaking the impact of my heavy lead-soled shoes, I felt the crunching of soft rotten wood and kept right on going. I was afraid that I might rip open the diving dress on a stray deck nail or on a sharp splinter, but before I had much time to worry, my feet struck something solid, and I came to a standstill. I was almost up to my hips through the rotten deck.

Laboriously pulling my feet back out of the hole in the deck through which I had fallen, I found it almost crumbly in parts. The deck timbers beneath it, however, seemed to be solid so that I could safely walk around without a complete collapse. I reached back down into the hole I had made to see on what I had landed and, to my rather pleasant surprise, found that it was beer—case after case, as far as I could reach, of bottled brew.

The *Tarpon*, we knew, was carrying approximately six thousand cases of Regal Beer. Although this would be secondary in salvage value to the contents of the safe, here it was and here, also, I was. It seemed wise to send up a few cases, if for no other reason than to boost the morale of the boys up on deck. For this purpose, what could be more appropriate? It was ice-cold beer from the depths.

I signaled for a net and then plodded over to the cargo hatch, choosing to load from the natural opening rather than through my accidentally made hole. Dropping down into the hold, I found myself standing on a solid floor of beer cases.

Bitterly remembering the lesson of the *John Dwight*, when I so diligently sent up the basket loads of whiskey only to later learn that they were just salt water, I resolved this time to take more caution. Still waiting for the cargo net, I selected representative bottles here and there out of their cases and held them up to the light. Slightly shaking them, I could see carbonation bubbles forming all of the way through. They were good!

The net arrived, and I loaded case after case after case into it until it was as full as it could be and then gave the line a prearranged tug to be hauled in. Rather than spend the rest of the dive just loading beer, I decided it would be best to start looking for the safe right away, since that was our primary objective. There was plenty of refreshment in the net now to last our gang on deck for altogether too long a time.

Crawling out of the hold, I walked aft toward the portion of the ship where, according to our pictures, the captain's cabin should be.

There *was* no captain's cabin. It was clear from the broken stubble of the wreckage that the entire superstructure had been swept away and was lying in a huge pile on the port side of the hull. Looking over the edge, I saw a sickening array of twisted girders, splintered wood, and miscellaneous junk. This was going to be no minor job.

I dropped over the side of the ship into the debris and started crawling around. There was almost an endless array of rusty bedsprings, soggy mattresses, rotten blankets, ship's furniture, broken glass, and corroded office equipment.

Over the lifeline came three long pulls. My time was up. I signaled back the answer—two pulls—"No." The diver always makes the final decision. For one hour, he is king. It is almost impossible for the boys up topsides to appreciate under what circumstances he might be down there. And thus, it is law that they may not pull him up until they receive instructions to do so.

One physiological phenomenon about a drunk, 'tis said, is that he is unaware of the state of his own inebriation. Thus, he will vigorously protest, "I'm aw' right," until the moment that he falls flat on his face. Perhaps we suffer a similar state of intoxication under the somewhat parallel effects of high-pressure air. Whatever its medical explanation, it is a certainty that a man's judgment is decidedly reduced under these conditions.

I forgot all about my long past deadline for return to the surface. Blazing in my oxygen-fired mind was the vision of that safe, probably just under the next piece of wreckage. I dug on.

Those three-jerk signals kept coming. I was, it seemed, even a little irritated at their persistence. I had to drop everything and signal back, "No!"—otherwise, they would assume that I was unconscious and just drag me away. What in the world was wrong with those fellows up there?

I was getting tired. I stopped to survey my work. Working by hand, as was my only choice, it was obvious that ahead lay a formidable job.

A flash flew across my mind in that moment of discouragement. I must be far overtime. I was suddenly as sober as I could ever be. I checked my lines for a clear run up out of the wreckage. They seemed to be OK as far as this pile was concerned, but high over my head they disappeared over the rail of the huge hull. There was no time to waste. I sent out my three long pulls, which they had been requesting for so long.

The answer came back immediately, and momentarily I felt my lifeline draw up tight about my waist. Soon, I felt the wreckage disappear from under my feet, and the slime-covered hull seemed to start easing by my window relatively downward. In a moment, the massive rail appeared, and I doubled up to an "akimbo position" to be prepared to kick myself clear of it as I was dragged by. Soon it disappeared beneath me, and I was free of the ship, again going through the blue-green void of water on my way back to the world I had left behind.

It is almost impossible for a diver to tell whether he is moving upward, downward, or standing still under these conditions, since there is nothing against which he can judge his relative movement. There are indirect observations, however, which over a period of time will allow him to observe his progress. For example, in going up, the pleasant increasing warmth of the water, the decided increase in intensity of the light, the corresponding brighter color of the water, the loud gurgling of the bubbles as the exhaust valve attempts to pass the rapidly expanding air, or the feeling of expansion of the dress from the body all indicate the rise upward.

A minute passed, and I noticed none of these well-familiar changes. I violently wiggled to determine whether I was within visual distance of the wreck but could see nothing in any direction. A jellyfish, like a huge glob of almost transparent gelato, lazily drifted by almost horizontally, making it obvious that I was standing still. My first stop for decompression would be at thirty feet; from the color of the water, I was far deeper than that. Something must be wrong. Groping above the helmet, I felt for my lifeline. It was taut as a steel bar. I tried to pull myself up a few inches on it, but with every ounce of strength that I could summon, I could not equal the terrific tension. The entire crew must be leaning on it.

My dress was partially distended, and I was not heavy. There was very little water inside of the dress, perhaps only a cupful in my right leg from the slight trickle at one of the breastplate studs. I must be stuck.

I made another violent squirm to find my air hose. Waving my arms frantically above my head, it soon was obvious that it was not there. The air was still pouring in, so it must have been still connected someplace. My predicament was beginning to be clear.

I soon found my hose with my left foot. It, too, was taut as a steel bar. Stretching almost vertically downward.

The hose was apparently fouled on the wreck. I was someplace in mid-water. I could not possibly get a signal over the lines, both being tense like the strings on a banjo. The boys on deck obviously had no way of knowing my predicament—I was just heavy. Worst of all, I was far overtime—and still soaking up nitrogen. There, I knew, would be hell to pay—if I ever got up to have a chance to pay it.

40

PAY AS YOU LEAVE

Suddenly, I found myself falling.

The water was blackening in front of my window. The dress was squeezing in on my body; the bowline, which had been cutting in under my arms, was now slack. I groped in space to grab something, but there was nothing to grab.

Just as suddenly, the lifeline drew up taut, and my fall was arrested. Observing the same signs in reverse, I soon found that I was going up again. As the line under my arms again grew taut, I felt down for the hose and realized that it was again tight and had stopped my upward progress.

A few seconds later, I again found myself falling and then, in a moment, being pulled up. Down I went—up I went—down—up—with as much to say about it as a yo-yo on a string. Obviously, they were trying to work the hose loose from the snag. As I later found, they had, by noting the great length of hose that was out, ascertained that it must be fouled on the wreck, there being enough overboard to reach all of the way down and half my way back. It was very comforting, in this lonesome world in which I was helplessly trapped, to realize that they were trying to help me.

Finally, they hauled me up to my maximum height and then just held me there. As I was wondering what would be next on their minds, suddenly I found myself falling again. Apparently, they were thinking, too, for after a drop of a few feet, I felt some tension on the line, indicating that they were somewhat retarding my drop. Cherry knew all about the theory of the squeeze and was lowering me as if it were his own chest that were at stake and yet fast enough not to unnecessarily waste any time. I hoped I would land on deck.

I didn't. Looming up in the blackening water was the rail of the wreck—except that I was on the wrong side of it. Another six feet and I could have grabbed it, but no frantic stroke that I could and did make seemed to span that vast gap. Again, the slime-covered hull raced by, this time going relatively upward. In a moment, up to meet my lead soles came a soft oozy mud bottom.

I reached for the lifeline. It was slack. I immediately signaled "OK"—one pull. Back immediately came the acknowledgment—one pull.

Climbing up that slimy side was now my first problem. At that depth, the air was trickling in slowly, so I shut down the exhaust valve and distended the dress. The folds gradually filled out, and I felt the fabric gently bellow out away from my body. Soon the weight disappeared off of my feet, and I saw the slimy hull move relatively downward.

Floating up to the level of the rail, I reached desperately for it but could not again quite touch it. A slight current sweeping over the wreck was floating me every second even farther away. There was absolutely no power that I had to bridge those few feet.

With the back of my head, I knocked the exhaust valve button on the inside of the helmet, releasing into the water a glob of air. As it scurried quickly upward, the corresponding loss in buoyancy of the diving suit upset the delicate balance I was maintaining in three-dimensional space and down I started.

Landing again on the bottom, I walked over to the wreck to try it again, this time planning to hug it as closely as possible. Starting out actually rubbing against it, I soon found myself at arm's length as I approached the rail. Realizing that it would, in a few seconds, be out of reach, I grasped frantically at patches of marine slime but all to no avail. As I approached the level of the rail, my extended hand was a good foot from it—but it might as well have been a mile as again I slowly drifted out through space.

A third time I tried it, this trip being crowned with success. As I was just about to duplicate the last failure, I managed to get a fingertip hold on a projecting barnacle. It was sharp, and I was conscious that I had cut my hand, but I would have to wait to inspect the damage. Clinging gingerly to the fragile hold, I floated myself in close enough to the hull so that I could catch the rail as I approached. Climbing over it, I stepped onto the ooze-covered deck and then paused to inspect my hand.

Leaking out from it oozed a purplish black fluid, curling out and disappearing into the tide current. It was blood, its typical red color almost nonexistent at this depth, over a hundred feet of water overhead effectively filtering out that end of the spectrum.

I pulled my hose taut and discovered that it disappeared almost horizontally into the blue void covering the wreck, indicating that it was fouled at some point aft of me on deck. I trudged along toward this place and soon saw looming up the explanation of all of my difficulties. On a piece of the protruding wreckage of the cabin, the rubber tube was wound in a bad snarl.

It would only be a matter of a few seconds before it would again be clear. As I stepped up to the tangle, the three jerks again came over the lifeline, indicating that the boys up topsides were concerned over what to do. I decided that I would just slip the turns off of the snag before answering, and then I could give them the "Haul me up."

Apparently, they were quite excited up there, assuming that I was thoroughly dead by this time even though I had answered all previous orders for any other reason—but not getting an immediate answer, they decided, in view of the difficulties we had been having, to haul me up. The line suddenly tightened up around my waist and up I started, helpless now to grab anything or to get a signal over the line that was hoisting me. I had several turns unsnapped and hoped that the rest could now pull themselves off.

Holding my hose with one hand, at about the same level in the water, I felt it tighten up, again leading vertically downward. The boys put a terrific tension on the line but to no avail. Realizing that they had been too hasty, I now felt them dropping me again.

Of all the disgusting things that could now happen, the worst occurred. Just an arm's length out of reach appeared the rail—and again, I was on the outboard side of the wreck. A moment later, my feet were in the soft oozy mud bottom, now to start all over again.

This time, I succeeded in floating directly up to the rail, again following my hose up to the point where it had fouled. I untangled the snarl, checked all of my lines for a clear run to the surface, and then signaled the three long pulls to the boys up topsides.

Up I went—and then my troubles were over.

At least for a moment.

I had made previous arrangements on extended decompression schedules in the event of just such an overlong dive as this. At thirty feet, I dangled for fifteen minutes, which seemed like an eternity after the hours I had already spent in that lonesome world. That was nothing, I soon found out, as the forty-five-minute period dragged on at twenty feet. Here, however, I had the first feeling that there actually was a world up there that I had left behind as I glimpsed the deep red keel of the *Arna* a couple of fathoms above my head. She was vigorously pitching and rolling from this fish-eye view, and I imagined that the wind had picked up considerably in these hours that I had been away.

It almost seemed like going home to finally find myself moving up to the last stage at ten feet. This, I knew, would be the longest, but there at my window level was the massive underbody of our boat. Behind that barnacle- and slime-covered copper-painted exterior were some warm dry clothes, my bunk, hot soup, friends, a radio, and all of the material things that made man's biological status just a little bit better than that of a fish.

The hour did finally pass, and I felt them pull me up through the last ten feet. There, suddenly in front of me, were the green rungs of the sturdy diving ladder. I grabbed one of them and started up, my eyes squinting in the bright light. As the helmet came off, I found myself peering into a circle of anxious faces. "What in the hell happened?" they asked.

"My hose was fouled on the wreck," I said, "but what happened up here?"

I could hardly see the deck of the *Arna* for beer foam. I couldn't make up my mind whether to laugh or to cry. It looked like a bubble bath from the set of a Hollywood movie.

There was one face missing in my little audience—our cook. Getting up, I saw him on the other side of the cabin coaming flat on his back on deck, soaked with beer from head to foot, his heavy jowls protruding from the sea of froth.

"Those bottles were spouting like a whale when they came out of the water," explained Cherry.

I had checked them on the bottom, and they were all good. Apparently, the discrepancy was to be explained in the following manner.

Down in the cargo hold of the *Tarpon*, there had been pressure on both sides of the bottle caps—the carbonation pressure of the beer on the inside and the pressure of the water on the outside—about balancing each other. In two years of submersion in seawater, a galvanic action had occurred, the caps apparently having been partially eaten away. As the bottles came up toward the surface, the pressure fell off externally as it stayed the same or increased (from the shaking up) internally. At about twenty feet, they had started to "blow their tops" so vigorously that the gushing foam allowed no salt water to run back into the bottle. Thus, on deck, there was ice-cold beer, opened and ready to drink in any quantity, but with no possibility of commercial resale.

I was hungry as a horse. I decided to go below and cook up a mess of ham and eggs and grits, our cook obviously being incapacitated for the rest of that day. I soon discovered that we were fresh out of both ham and eggs, but someone had caught a big red snapper, and so I changed my menu to fish and grits.

In the meantime, Cherry was slipping into the diving suit. He had never dived on a wreck before but was a very fine sponge diver. He was one of the few Americans in the business, and I was confident that he could get along nicely down there, being thoroughly at home in the water.

While my snapper was cooking, I went up on deck to give Cherry the details of my work on the wreck so that he could go on from where I left off. I was so hungry that I was miserable as I described the wreckage of the cabins and my theory as to the location of the safe. As they dropped the helmet down over his head, I hastened back below to enjoy my snapper dinner. I must have cooked a pound and a half of the fish, all that the frying pan would hold, not with the intention of eating it all that meal, but so that anyone who wanted could nibble on it later or make sandwiches as he chose.

I ate a large serving of the fish, about half a pound, with a proportionate amount of grits. Supposedly through with my dinner, I went back up on deck. I was just as hungry as before.

I went below again and started nibbling on bread and fish. Soon, I had finished up the latter and all of the grits. Before long, I had opened up a can of corn, fried up a few potatoes, eaten the best part of a loaf of bread, and cleaned up a number of odds and ends left over in the gallery. "I must be through by now," I thought, and again I appeared up on deck. Cherry's bubbles were fuming up off the starboard quarter, and all was going well.

It seemed as if I were hungrier than ever. Not knowing what else I could eat, I just stood there, leaning on the companion slide, wondering what was wrong. Suddenly, I felt a sharp pang in my stomach. It was now obvious that there was something wrong, that my terrific appetite had not been a normal craving for food but an artificial hunger. A moment later, I felt a "green feeling" coming over me. With not a split second to spare, I arrived at the appropriate place at the rail of the *Arna* and said snapper soon was returned to his native Gulf of Mexico.

I must be seasick, I thought, as I stared out over the trackless Gulf, *but how could I be seasick?* I continued to wonder. We were now listlessly riding to slack anchor lines on a flat, calm sea.

Were the fish poisoned? Was this a plot to try to kill me just as Davy Jones was about to yield his jealously guarded gold? My mind seemed to be deliriously stumbling through a kaleidoscope of frightening thoughts. I felt as if my body were a ghastly green or lavender color. I was afraid I was going to die. I was afraid that I was not going to die.

I stood up. I was extremely weak. "Mister, you are sick," I said to myself.

My legs crumpled beneath me. I collapsed in a heap on deck; I tried to get up. My legs wouldn't move. They were paralyzed.

I now knew what was wrong. I had the bends.

There was a moldy blanket a few feet away. I reached for it and crushed it up into a ball to make an improvised pillow for my head. The deck was so hard.

There was a splitting pain running through my right arm. That slight effort had seemed to intensify it to unbearable proportions. Suddenly, the pain disappeared. I tried to tuck the blanket under my head. Nothing happened. My arm wouldn't work. It was paralyzed.

I tried my left arm. The same sequence happened—pain followed by paralysis. I tried desperately to move my legs again. They wouldn't move.

It was now time to ask for help. I called one of the boys and suggested that he give Cherry his signal to come up. There was only one cure in the world for the bends and that was to go back down under pressure again.

A moment later, I heard the tender sing out, "We're hauling him up." Apparently, Cherry had received the signal and returned instructions to "Haul away."

I looked out over the placid, listless water. There seemed to be a low fog drifting lazily along. I looked at our taffrail. I couldn't seem to focus clearly on it. It was so hazy. The fog seemed to be getting much thicker. The light was rapidly fading. It must be late. I could hardly see anything now. Suddenly, I remembered that actually it was early morning. I was very frightened. I frantically blinked my eyes. I couldn't seem to see anything. I was panic-stricken. I was blind.

Worrying wouldn't help any. Cherry was on his way up. He still had some decompression time ahead of him. It would be a while before the suit would be ready. I might as well relax and be patient.

My eyes seemed to be getting a little better. I could barely see light again. I could feel that they were open. I now could barely discern the rail of the boat. It was getting clearer and clearer. I could see the deck. I could now dimly see the mast. Everything would soon be all right.

I tried hard to focus on the rail. It still seemed so dark. There were a few fish swimming under it. The deck seemed to be covered with a deep green moss.

The veil seemed to be rapidly falling. Suddenly my mind was crystal clear. What were those fish doing up here? Where did that slime come from on the deck? What was that metal enclosure around my head?

I was on the deck of the sunken *Tarpon*.

41

THE BENDS

I stood up. I felt fine.

"Hey!" I exclaimed to myself. "What am I doing standing up. I'm supposed to be paralyzed!"

My arms were both back. My legs felt strong. The pressure had completely cured me.

I started to walk over toward the rail to pick up some clue as to where I was on the wreck. I took a step and stumbled. I was tangled up with some huge hunk of something.

Kneeling down, I tried to feel out what it was. Bending forward, I suddenly saw in front of my window the bright galvanized ring of an anchor. Studying it over, it was immediately obvious that this was in decided contrast to the rest of the slime-covered wreckage.

It was our own anchor. I was tied fast to it. It was now all very clear. They had fastened me to it and had thrown the combination overboard while I was unconscious. It had been necessary to add this extra weight since I had not been in a position to operate my exhaust valve and thus could not blow out the excess incoming buoyant air.

I felt strong as an ox and carefree as a drunk. I walked around on the sunken deck for a few minutes, exploring a part that I had not seen before. I was in no hurry to get back, rationalizing that a few extra minutes would be advisable to get my bends shaken down. However, every minute was just adding to the total damage, so soon I gave up the signal, "Haul me up."

The response was immediate, and in a few seconds, I felt the deck melt away beneath my feet, and I was on my way back. The water rapidly brightened, but I knew that there would be another long decompression before seeing the world of air again. My progress suddenly stopped, and I guessed that I was now at my first decompression stage, thirty feet under the surface.

The thirty-foot stop seemed endless. Up I went at the end of it to twenty feet. It seemed as if I spent an eternity there. Finally, up I went to the last stage at ten feet, the longest of the three, but somewhat less lonesome because of the presence of the nearby underbody of the boat. I was so happy I almost felt like crying when I finally felt myself moving upward again and momentarily found my window face-to-face with the sturdy rungs of the diving ladder.

I climbed up a few steps. They took my heavy lead weights off and then the helmet, as I stood waist-even with the gunwale of the boat. Now light, I climbed up on deck and prepared to get out of the suit.

Suddenly, I thought a streak of lightning had hit me. A blinding pain raged through my legs. A moment later, the same pain flowed through my arms, and I felt as if a steel wedge were being split through the very marrow of my bones.

My agony was short-lived. I crumpled on deck. My legs were again paralyzed. Tears were running down my face. I couldn't stand the pain in my arms. That, too, seemed to ease. I seemed to feel much more comfortable. I opened my eyes to again face the boys, who I knew were intently watching me. I couldn't see them. What were those fish doing? There were hundreds of them grouped around staring at me. My mind was suddenly very clear.

I was again down on the wreck. There was that anchor again, tied to my waist.

I gave the three-pull signal. I would try it all over again. Up I went, this time to briefly stop at thirty, twenty, and ten feet. In half an hour, I was on the ladder.

They lifted my helmet off. "If it happens again," I requested, "please don't send me all the way down. Drop me down to thirty feet, and see if I can regain consciousness there. If you don't hear from me after five minutes, start dropping me very slowly until you do. If—"

I couldn't understand. There was nobody there. I was just going to explain further.

I was in the diving suit, hanging someplace in the clear blue water. I couldn't be very deep—the light was so bright. I wasn't moving. I looked through the upper window and saw the silhouetted hull of the *Arna* against the flashing silver surface. I must be at thirty feet.

My bends were gone. I grabbed the hose to give the "OK" signal. The answer came promptly and methodically back, and soon I felt myself rising up to what I thought must be the twenty-foot stage.

Half an hour later, I was again struggling up the diving ladder, feeling fine but somewhat skeptical about how long it would last.

Again came the streak of blinding pain. This time, I did not lose consciousness. I was making progress, but not so sure that I was pleased about it. I couldn't stand that pain. A team of horses seemed to be harnessed to each arm and to each leg, all pulling as hard as they could in as many directions.

I felt myself going down again. This time they planned to drop me only to twenty feet.

I stopped at that depth and just waited. Not over half a minute later, the magic wand was again waved. The pain suddenly seemed to just dissolve into nothing.

I gave the signal. Up I went to ten feet. I hung there for thirty minutes. I had no trace of the bends! I wondered if I could ever safely get through that measly last ten feet of water. That half hour was another eternity. Although I had no trace of pain, I was about as thoroughly miserable as it was possible to get. The diving suit had many hours before taken on a very slight leak, and the small incessant drip of water had been running down my back. The water was not cold. To a swimmer, it would have been ideal for a refreshing swim. To me, almost to the point of physical exhaustion, it was an irritating medium which seemed to be slowly draining the last calorie of heat out of my body. My blood, it seemed, was gradually approaching the temperature of the Gulf of Mexico. My teeth were chattering violently.

Finally, the eternity came to an end. They pulled me up. I got on the ladder. CRASH!—and the streak of lightning struck again.

Down again—up again—down again—up again—and thus, on it went far into the night. It was consoling to note that each time that I attempted to leave the water, the bends struck me with less intensity. Each time, I had discovered relief at a lesser depth. I was now vacillating between the level of ten feet and the surface. I couldn't seem to conquer those last few feet.

Again, I was up on the ladder. The moon was peeking out occasionally from behind clumps of clouds, which were racing madly by. The wind was brisk. The seas were responding to it with a short chop. The *Arna* was dancing discontentedly to her hawser in the inky darkness.

The bends, as they struck again, had resolved themselves down to mere pain in my arms and legs. Although there was no actual paralysis, my skeletal muscles were so exhausted and were in such agony that they were virtually useless to me. I could move my arm or leg a little bit, but the thought of using it for any useful purpose seemed rather remote.

The boys on deck were all exhausted. They seemed extremely irritable, with the incessant motion of the boat serving to exaggerate their uneasiness. Each time that I had come up, they asked with much concern how much longer I thought it would be.

My progress had apparently been leveling off. I had for some time been finding relief at ten feet but was now apparently noticing little improvement each time the symptoms again struck me on the diving ladder. I was so miserable that I didn't much care what happened. The sloshing water in my suit seemed almost worse than the blinding pain.

I decided to try staying up on deck. If things got much worse, I could go back again later with my dry underwear on and in a dry dress.

They laid me down on my belly on deck. All hands turned to, to hoist anchor. Five minutes later, I felt the wind shift to our starboard beam and heard the run of the water along the hull as the *Arna* took on full speed ahead toward the port of Panama City.

It was three o'clock in the morning. I had started to dive at ten o'clock the previous morning. Except for the very brief interims where I had unsuccessfully tried to come out, I had been in the water for seventeen hours.

With the boat now on her course, the boys returned to get me out of the suit. They stripped me down to my goose-pimpled skin, dried me off, and poured me into a suit of dry, warm diving underwear. This was a sensation which heaven would hesitate to rival. In a few minutes, they put me in my bunk, covered me up with blankets, and then, except for the helmsman and engineer, turned in themselves for a few winks.

I couldn't sleep, the pain in my arms and legs being too severe for that, but it was comparatively pleasant to just lie there and feel the throb of the diesel as it faithfully pushed us on toward dry land.

The first dim streamers of dawn were appearing in the east as we nudged into the city dock at Panama City. Cherry and I had decided that the best thing to do would be for me to go to the hospital. Here I could have a warm dry mattress, suitable nourishment, and, above everything, some morphine to sedate the severe pain in my extremities.

Cherry jumped on the dock and hurried on into the town to search for a late taxicab. In a few minutes, he returned with his find, and the boys helped me off the boat, down the pier, and into the waiting car.

I couldn't yet walk but could partially support myself by leaning on someone. Cherry and I arrived at the front door of the hospital. It was locked.

Groping, he found a bell and pressed.

In a few minutes, the door opened. A nurse, typically bedecked in her white uniform, appeared.

Cherry is difficult to understand under the most favorable conditions. He stutters. To cover this up, he hurries over his words at a pace just a little faster than his vocal mechanism will operate. It had taken us many months to learn to understand him. However, in the presence of strangers, his difficulties seem to increase a thousandfold.

He started babbling a blue streak, which must have been as intelligible as a record running backward. The nurse, totally failing to understand, made a quick survey of our figures. Cherry was bedecked in a grease-begrimed pair of pants and shirt that would have properly been discarded overboard by our engineer as filthy rags. He had on a battered old hat that was intended only to be worn, without exception, in the middle of the Gulf of Mexico. I was wearing only my long woolen underwear and an old torn pair of pants. I had, apparently, lost my shirt and shoes.

The door slammed in our faces. We heard the bolt being slid in place. After several jingles, she again appeared. Cherry started again, this time even faster, to explain our request.

I didn't feel like talking but realized that some translation would be necessary.

"I have the bends," I explained. "I would like some morphine."

"You have *what*?" she asked.

"The bends," I repeated. "It's a diver's disease. Perhaps you might call it caisson disease."

She just stared at us. Obviously, she still thought that we were either drunk or crazy.

"I am in pain," I explained. "I must have some morphine."

"I'll have to see the doctor," she finally said. "What was it you said you had?"

I repeated "the bends" to her. The door closed. The bolt again slid into the locked position.

In a few minutes, we heard the lock opening again, and the door opened.

"What was it you said that you had?" she again asked.

She went back to the doctor with the elaborated explanation. Apparently, he had now consented to our admittance for she now opened the door and let us in. She led us to a room, turned on the light, and suggested that I lie down on what apparently was the X-ray table.

"If you will lie down, I'll give you the injection and then I'll get you a room."

She appeared in a minute with a syringe. As she slowly injected the drug, I felt the welcome numbing effect stealing over my body, reducing my pain to more and more bearable proportions.

I blinked my eyes. I must have nodded off. There was no one in the room. I wondered where the nurse had gone. I felt strangely rested. There was still much pain in my arms and legs, but the latter seemed strong enough to support my weight. I hobbled over to the door.

Opening it, I was immediately conscious that it was many hours later. There were several nurses, patients, and doctors walking by. In a window down the corridor, I noted the nearly overhead sun streaming in.

The nearest nurse stopped in her tracks. "Who are you?" she asked.

I could answer that question but could not give her much information about my official registration in the hospital. From the feeling of my back, apparently, I had slept on the metal X-ray table all night. There had been empty rooms available. Several people had come in that morning looking for me, but the hospital had no record of any such person. While we were speculating as to what had happened, in walked Cherry. I heard him saying, "—but I brought him in here myself late last night," as he saw me.

The hospital was very apologetic and immediately offered me a room. However, I felt much better, even though I could move my arms and legs only with much difficulty and even though the pain had certainly not vanished.

Although it was almost noon, the boys were anxious to get out again to the wreck. Cherry wondered if it would be all right if they sailed.

"Sure!" I said, "and I think I'll go along with you." If I was going to be in agony, I might as well do it on the boat as by myself on shore. It was a beautiful day, and diving conditions would be perfect.

We had lunch as we pushed our bows out to sea. As hungry as I now was, I found that my arms just weren't strong enough to lift the weight of the spoon all the way up to my mouth. Cherry had to feed me. This was the height of humiliation.

Our course was good, and we sailed almost over our wreck buoy. Anchors were soon laid. We were ready for diving.

I looked over the side far down into the mysterious clear blue-green water. It looked far different from the black depths of last night. I shouldn't have looked down into that water. It was too much.

"Think I'll take the first dive," I told Cherry. That water had worked magic for my pain yesterday. I wondered if the formula would work today.

It was quite a task getting into the gear with my paralyzed arm, the maneuvering and wriggling ordinarily being done by the diver. Once I was dressed, however, it was easy to roll overboard, and soon I found myself sinking rapidly down to the wreck.

In about a minute, my feet hit her slimy deck. I could feel the magic wand slowly waving over my head. I could feel, by some strange inner sense, my bends being crushed into oblivion by that tremendous water weight over my head. Suddenly, they were gone. I was strong as an ox. I had no trace of pain anyplace in my body.

Today, however, it would be a short dive. For thirty minutes, I cleared wreckage, and then felt the three pulls as we had previously agreed. I took 50 percent longer decompression than was standard for this stay on the bottom, not having the slightest desire to repeat yesterday's experience.

About ten minutes after I came out of the water, the old pains gradually started to settle in. Yesterday, they had struck like a bolt of lightning; today, they just grew gently. It seemed as if now they did not return to the intensity they had been this morning. Apparently, I had accomplished some good in today's dive.

Thereafter, I dove every day. The pressure seemed each time to offer relief; return to the surface brought a recurrence of the pain, but on each day, it seemed a little less. Soon, it was all history.

42

Topsy-Turvy World

One week later, I again found myself face to face with temptation. I was in a dimly lighted cabin of the *Tarpon*, already one flight down from the main deck. Should I venture down to the mysterious deck below? As I peered down through the companionway, all that I could see was a black void.

"A diver should never venture into any place where his lines do not have a clear run up to the surface," I recalled from my diving bible. This was good sound advice.

However, this was different. I was not diving with any makeshift, improvised, or unreliable diving rig. Up topsides, I had a fine heavy-duty diesel engine driving the multiple-cylinder air compressor, both in perfect mechanical condition. Just in case this invariably reliable machinery did break down, however, coupled into the line was a huge receiving tank filled with high-pressure air, which, with the flick of a valve, would supply me for all the time that it would take to get to the surface. And then, just in case both of these simultaneously failed me, there was a hand pump connected in, which, although it would not give me a surplus of air, would get me up to the surface. It was almost inconceivable, in short, that there would be any possibility of air failure.

I decided to try it. Down I went, my heavy lead shoes digging into the soft, worm-eaten, wooden treads of the steps. The handrail, which I fancifully imagined many a sailor had grasped as the ship had lurched in a heavy sea, felt spongy to my grasp. I gave it a little jerk to see just how solid it was. A hunk came out. I broke it in two like a long stick of Greek bread. It was rotten all of the way through and riddled with worm holes.

I remembered the heavy planks of the main deck had crumbled under my weight. I wondered how soon the entire ship was going to collapse.

I stepped out into the stygian darkness of the cabin before me. I could see nothing. I had a feeling of foreboding disaster. Two decks and 130 feet of water spread above my head, separating me from the world of air.

"Let's go," I said to myself. "You are just afraid of the dark!" Perhaps that was really true, and I had no real cause for alarm. I ventured on into the cabin, feeling out strange-shaped objects for what might be the ship's safe.

Moving as fast as I could in the inky water, I crashed helmet-on into a heavy wood stanchion. I felt it give before the combined weight of my suit and my body. There was then an eerie, indescribable, subconscious feeling of the collapse of the overhead deck. For that uncertain second, I felt myself crouching, turning away, and pulling my neck in like a turtle.

Cr-a-a-a-a-a-sh!

Down on my head came a mass of rotten ceiling, deck frames, and planking. Through the water came the sound of the reverberating crash of what must be the entire overhead structure of the wreck.

This mass falling down on the head of a man under ordinary conditions in air might have killed him instantly. Down here, it didn't even hurt. The wood was soft and spongy. Buoyed up by the water, it weighed only a little more than nothing. Retarded by the water, it could not fall very fast. It was little more dangerous than falling snow.

I attempted to stand up, pushing my helmet up through a maze of rubble and debris. Turning, I could no longer see the faint gleam of light that had marked the companionway down which I had come.

Mentally summarizing my situation, I quickly realized that something was wrong with my air. Some must still be coming through. I could hear the faint pulsing of the air compressor, probably four hundred feet away through the long narrow-bore passage of the hose. I could barely feel the incoming air blowing against my face. The air compressor must still be running, but it certainly wasn't delivering its usual volume. The customary blast was now just a gentle breath on my head. The hose must have been largely pinched off in the wreckage.

I attempted to turn to start what probably would be a laborious process of plowing my way through the wreckage. I knew I must follow my own hose and clear whatever had fallen on it.

I couldn't turn around. Something was holding me tight. I tried to turn the other way. Again, I was held. Moving up and down, back and forth, and gyrating, I soon realized that it was my own hose that was pinning my back to the wreckage. I succeeded in finding my lifeline and discovered that in it, I had several feet of slack.

I squirmed and squirmed and squirmed. However much I contorted myself, I couldn't seem to be able to pull out enough slack to turn around. With a desperate effort and a strained reaching of my arm, my right hand finally found the rubber tube. I grasped it and pulled as hard as I could. It was as immovable as if the Rock of Gibraltar were resting firmly on it.

Relaxing to consider my plight, I realized that I was panting violently. My contortions had consumed a lot of energy. Not enough air was coming through the constricted hose to provide for it. There were two things that I had better do immediately. The first would be to relax every muscle in my body as completely as I could for a few minutes. This would calm my breathing down and get my body in better adjustment with the meager air filtering into the helmet. The second would be to plan, coolly and with no trace of emotion, exactly what I should do.

From the limited amount of feeling that I could attempt in this position, tied up like a dog on an extremely short leash, I realized that my chances of getting over that pile of debris, even if I were free from my air hose, would be very slim. In the apparent collapse of the wreck, the possibility of the narrow companionway passage back to the deck above still being passable seemed very remote. The crash had apparently been behind me. As far as I could feel in front of me, there seemed to be fairly clear passage.

I was now completely relaxed but found that I was still panting hard. The amount of air trickling through the hose seemed to just about or perhaps not quite meet my basal requirements. The helmet was very foul. Whatever I did, I must do it very soon.

I decided that there was only one thing to do. I must cut my air hose.

I felt for my knife and found that it was in position at my side. I slipped it out of its sheathe and mentally reviewed the exact plan that I would follow after even this meager trickle was cut off. I would try for another exit from the cabin, confident that there must be another companionway at the after end.

Everything was rehearsed. I was ready to go.

First of all, I cut the lashing which held on my shoes. This would lighten my feet by about thirty pounds.

Next, I cut my air hose. A strange sensation crept over me. From now on I was strictly on my own. Lastly came the straps, which held on my diving weights. This would remove another hundred pounds and make me buoyant as a cork. My suit was dry as a bone, and I could enjoy its full buoyancy. As I cut the straps, I felt the heavy weights slide off my lap to the deck below. From there on, things happened quickly.

I started moving like a released jack-in-the-box. My legs, being light, instantly turned up and filled with air. I shot upward, feet first, and in a second felt my shoeless soles strike the partly collapsed overhead deck. I was now in a topsy-turvy world. I could walk even more naturally on the ceiling than a fly.

The law of gravity was now totally replaced by the law of buoyancy, my feet being pressed upward against the ceiling as firmly as they would be on the sidewalk during a stroll through the park. I could walk just as certainly and in just the same way as I had done in air ever since I was a baby.

This was no time, however, to revel in the curiosities of nature. I had a very limited number of minutes of consciousness left, and in those fleeting minutes, I must find a way to get out of the wreck. I walked along, moving away from the wreckage. Stumbling over a sharp metal object protruding upward like a bayonet, I realized that it must be a chandelier.

In a moment, I bumped into a wall. Following this along, I soon came to what I was looking for—a rectangle of light in the ceiling. This was the companion opening for which I had been looking.

Standing at the edge of it, I could see "down" into the comparatively bright room "below." Looking up, I saw the stairs "above" my head, which were normal course for passage between the two cabins but which would be of little service to me now. There was no time to act deliberately. I just jumped.

"Down" I went for a "drop" of about ten feet, my feet landing softly on the ceiling of the next cabin. I could now see quite well, comparatively, and, after a quick survey of the cabin, started off toward another rectangle of light which represented the passage of clear water to the surface above.

In a moment, I was standing at the edge of this rectangle. Looking "down," I could see the vast expanse of blue-green water "beneath" me and realized that this jump would take me all of the way up to the surface.

I now realized that I was essentially back to the world of air. The law of buoyancy would as certainly shoot me up to the surface as would the law of gravity plunge me down if I stepped off of the roof of a ten-story building.

I was panting violently. For the past few minutes, it had seemed as if I was staying conscious by sheer willpower. I felt very unstable and very uncertain that I could retain my sense more than a few seconds longer. My first real sense of fear seemed to now shoot through me—fear for something that was beyond control. Would they see me at the surface?

I would go up feet first. Up there, I would float like a buoy, my head and shoulders being vertically downward. There would be no conceivable way that I could right myself any more than one of the lead-based saltcellars could stand by itself, upside down. My feet were light; my helmet was heavy. There would be no possible way that I could call the boys or attract their attention.

The routine on deck when a diver is down flashed through my mind. My tender was probably sitting there, staring out at the water and my frothing bubbles with his mind a million miles away. Those bubbles would still be tumbling to the surface from my severed hose, revealing to him where I was supposed to be.

The rest of the boys were probably belowdecks. The cook might be preparing lunch in the galley. The engineer was probably tinkering with his tools or equipment down in the engine room. The rest of the crew, on call the instant they were needed, were probably busily engaged in a spirited craps game down in the main cabin.

I was frightened. If they didn't happen to see my feet sticking out when I came up, within a few minutes I would drift away and be lost from sight in a dancing pattern of waves. Once I jumped, I was helpless.

There was little preparation I could make. My fate was now just pure chance. I jumped. I started moving toward the surface with terrific speed. The fear of the lonesome wastes of the Gulf of Mexico now seemed to grow to tremendous proportions. My heart was pounding harder and harder; my panting was growing more and more desperate; my head was pounding like a trip-hammer. They all seemed to be in perfect rhythm—a dull, heavy oppressive thump-thump-thump.

My eyes seemed to be closed. I tried to open them to see how I was progressing. I could tell my depth by the color of the water. Who were all those people bending over me?

They were talking. I could hear someone say, "He's OK—his eyes are opening." I suddenly realized that I was on deck. I could feel the heavy throb of our diesel engine. My head seemed to be throbbing in time to it. The light was hurting my eyes.

I sat up. Except for a headache big enough for a hippopotamus, everything was all right.

The boys tried to figure it out. No one had seen me come up. No one knew how long I had been floating at the surface. The cook, waiting for the stove to heat up, had been leaning in the companionway, idly staring out over the water. We were many miles offshore. "What's that?" he called to the tender, pointing at a strange-shaped floating object barely discernable several hundred feet off our port quarter.

The next day, I didn't dive. I didn't eat. I didn't even want to smell diesel oil. I slept in a bed in town and spent the day eating aspirin and wrapping my head in wet towels. The weather was not good, and the boat didn't go out anyway.

The following day, we were out on the wreck again. I followed my air hose down to where it was snarled and cut it off from the upper end, salvaging probably all except about twenty feet of it.

Day after day, we continued our search of the wreck. We salvaged everything that we could tear loose. She was carrying a mixed cargo and, piece by piece, we went through it to sort out anything that might have resale value. Although the ship had only been down a year, much of the cargo had been damaged or spoiled by the salt water. We were pleased to note that we were not only making expenses on the job but were actually showing a little profit for our efforts. But where was that safe?

We were about ready to give up. There was not another part of the wreck we had not thoroughly searched. The safe was just not there.

It was the last day of diving. I was down on the wreck and frankly puzzled as to what more I could do. I sent up a few more cases of beer and cursed, knowing that they would blow their tops on their way up and thus ruin their possibility for resale.

I walked along the port rail, staring down into the deep water over the side. A few huge turtles were scurrying along, and I watched their armored bodies as they glided under me. I was leaning far over the rail. Suddenly, the structure gave way. I found myself falling. I was not too concerned about it, however, since I was wearing the light leads. I could control my fall by admitting a blast of air and increasing my buoyancy.

Settling slowly down, I studied the side of the huge hulk as it passed by my window. Suddenly, my feet were stopped, and I realized that I was on the bottom.

Looking down, I saw that I was not standing on sand, but rather on a huge pile of wreckage. There were bedsprings, broken-up picture frames, and splintered timbers. This, apparently, was the wreck of the ship's superstructure.

Perhaps here would be the safe! It had been my contention from the very beginning that it would have been in this topmost portion of the vessel, but the difficulty had been thrown off either by the crashing seas as she had sunk or by a lurch as the huge hulk had settled on the bottom.

I started digging in the debris. It was not long before I found that my theory had been correct.

There was the safe!

By prearranged signal, down came a huge hawser. I slung it around the large box in such a way that by no unforeseen lurch of the surface vessel could she get away.

With the safe firmly fastened, I gave three tugs on the line and, in a few moments, felt it draw up taut. Now on the windlass, the tension straightened that manila out as stiff as a steel bar. Then the box moved, first an imperceptible part of an inch, then a lurch of seven inches, and finally it reluctantly lifted out of the rotten timbers which had been its bed for over a year.

I stood there and watched it slowly move upward, making sure that it didn't snag on a protruding portion of the wreck. Now it was clear for this long trip up to the surface. I checked my own lines and prepared for my own return to the world of air.

I had almost an hour of decompression. That was a long hour. I just hung there, imagining that the boys had my prize solidly resting on deck and had, by this time, knocked off the combination lock with our big sledge hammer. They probably already had all of the money out and were busy counting it. It was a shame that I had to spend that exciting moment just hanging on a rope a few feet under the boat's keel.

When my last ten-foot stage was complete, they pulled me up, and I eagerly reached for the diving ladder to get up and see what luck we had had.

On the top rung, they pulled off my helmet and weights. I stood up and looked toward the bow. I could see no safe.

"Where is it?" I asked, afraid that something had gone wrong.

"We have it," the engineer explained, "except that it is so heavy we can't lift it out of the water."

Taking off my diving shoes, I walked forward, still in dress and breast-plate. Looking over the bow, I noted that Cherry was in swimming, hanging on to a huge slime- and barnacle-covered box just under the surface. He was trying to sling another chain around her just to make certain that the rusted sharp edges or the barnacle coating didn't chafe through our manila lines. But there she was!

The safe had been quite a load, just hauling it up this far. But now, the problem of lifting it out of the water presented serious difficulties. A great part of its weight was still being supported by the buoyancy of the water. As our bow rose and fell in the passing waves, I could see how dependent we were on the buoyant lift; occasionally in the trough of the sea, as much as half of the volume of the safe was suddenly out of water. The added weight on our bow was violently protested by our diving vessel as every timber in her hull seemed to cry for mercy.

There was only one thing that we could do, as torturous as it would be. We would have to leave the safe submerged where it was and, in this position, tow it underwater twenty-five miles to the nearest port. Here, we recalled, we could find a heavy derrick, which could lift our entire vessel out of the water if necessary.

The huge mass dragging in the water slowed us down to not much more than swimming speed. On deck, as we labored along, there was no lack of talk, speculating as to what luck we would have. We had the safe tied up so securely that we could have ridden through a hurricane if necessary and not lost it—or if we did, our boat would have gone down with it.

The next day, the crane was in operation, and they picked the huge weight out of the water. We moved our bow underneath, and then they

"Buckets of money" are retrieved from the safe of a sunken ship. Nohl's ace diver, Maurice Gay, is removing the last batch of bills from the safe; in his right hand is a security box. US currency is not damaged by submersion in salt water—this money had been submerged in 120 feet of water for over a year.

gingerly dropped it down on our forward deck. The entire town was out to see the sight—the docks were packed three-deep with people.

However, if they expected that we were going to open it before their eyes, they were in for a disappointment. We had long ago decided that if we did find that safe that there would be only three people in the world who would ever know what was in it. It had been lost and abandoned at sea. By international law, it was our privilege to salvage it, and unless some rare suit were won against us for a portion of it to be returned, it was our privilege to keep the contents. We were under no obligation to report to anyone what we found.

The steamship company, we had been pleased to note, had been in confusion as to how much money actually was on board the ship, and even better, exactly where it had been put for safekeeping. Captain Barrow had been a cautious man and had sometimes hidden part of the money in other places than the safe, such as in a locked desk drawer or in his cabin for protection in case of piracy at sea.

We opened the safe. The boys helped us burrow through the heavy steel and concrete walls. But just before the door actually swung open, we sailed out into the bay, away from all inquisitive eyes.

There are only three people who know what was in that safe. Cherry was one. He won't tell. I was another. Cherry and I had, the day we sailed for the wreck, drawn up a solemn covenant that we would never tell.

Captain Barrow was the third. He went down with his ship.

43

The Pot of Gold

I dropped out of DESCO in 1940 because I sincerely felt that it would never amount to anything. At the time that I left, our combined gross sales for the two years of the company's life made a grand total of zero dollars.

I might have been right under ordinary circumstances. Although war had started in Europe, I, like millions of other Americans, was a firm believer in isolationism and I did not believe that we would ever enter the conflict.

Then came Pearl Harbor. Overnight, we became engaged in the greatest war that the world had ever known.

As the might of the great German submarine fleet was now directed at our shipping, aimed at both our merchant and our fighting fleet, it was apparent that there would be a demand for diving equipment such as the world had never before seen. A good part of our Pacific fleet was lying on the muddy bottom of Pearl Harbor on the very day that we declared war. As the fighting progressed, there developed such mammoth diving jobs as the raising of the scuttled fleet at Massawa in the Red Sea, the raising of the giant liner *Normandie*, and the countless other vitally important jobs around the entire world wherever there was an enemy engagement at sea. New war and merchant ships were soon sliding down the ways on a production line basis at a staggering rate. Each one of these vessels would now be equipped with diving equipment so that she could, in an emergency, in a far corner of the world, put a man down and repair the underwater damage of a battle at sea or service the propellers, rudder, and underbody of the ship.

Suddenly, the government required diving equipment in quantities that the world had never before known. I had, when I was with the infant DESCO, gone to Washington and made a vigorous attempt to sell them our wares but hadn't come back with an order for a nickel's worth of equipment. However, the name of DESCO and our attractive literature was now on the official government purchasing files.

The story would now be different than it had been in peacetime years. Uncle Sam would now finance, in part or in full, these contracts. It would require a minimum of investment on the part of Kuehn, the backer of our company.

DESCO sprang to life. To make a long story short, by the end of the war, she had become the largest manufacturer of diving equipment in the world. Although the exact production figures were shrouded in wartime secrecy, she may have been now larger than all the others in the world put together. I was out, by my own volition, of the company that "would never amount to anything."

When the war came to an end, it was understood that there would be no more government contracts for a long time—certainly no more really substantial ones until such time as we again, at that time an inconceivable thought, went to war. There was still the commercial diver's market, however, which had grown to heretofore undreamed-of magnitude. During the war, the Navy Department had trained tens of thousands of divers—had given them the finest diving training that was available in the world. A good part of these men wanted to enter professional diving as soon as they got out of the service and would certainly need equipment.

However, it was soon apparent that in spite of the greatly increased number of divers, DESCO would be selling them little equipment for a while, too. Millions and millions of dollars of DESCO equipment was now being dumped onto the surplus market. It was selling at a price that was far less, in most cases, than our lowest possible direct cash manufacturing costs.

Norman Kuehn and Jack Browne could see very clearly the handwriting on the wall. There was a group of men in Milwaukee headed by Alfred Dorst who were interested in the entire stock of the company at the right price. They together owned a metal-stamping company and were involved in a number of real estate investments. Their thought was that in spite of the gloomy-looking immediate future, the name DESCO was now tops in

At one time, Nohl believed DESCO "would never amount to anything." But by the end of World War II, it had become the world's largest diving equipment manufacturer.

Rows of diving helmets manufactured by DESCO.

diving equipment throughout the world. Eventually the surplus would disappear, eventually the navy would be replacing its equipment, and sooner or later there undoubtedly would be new technological advances to open an entire new market.

The field of water sports was suddenly growing by leaps and bounds. Although the merchandising of this equipment would be through quite different channels, by way of jobbers to the sporting goods stores, the company had the manufacturing set-up to make such apparatus. Soon DESCO offered such items as an underwater swimming mask, swimming goggles, swim tails for the feet, nose clips, and earplugs.

Unsuspecting, step by step, they even evolved to equipment that was made to be used on top of the water. The link in this drastic change in policy was a device known as the "Search Board." It was essentially like the Hawaiian surfboard, so well popularized at Waikiki Beach, except that the DESCO device had a glass waterbox cleverly built into it so that the rider could study and enjoy the beauties of the submarine world over which he was passing. It was a successful cross between a diving mask and a boat. With the acceptance of the Search Board, it was an easy step to the Aquaplane, which is a type of board that is towed behind a boat. This completed the transition, since the Aquaplane had no possible connection with the world underwater (except when people like myself tried to stand up on it and soon found themselves pitched head first at high speed down into the submarine world). From the Aquaplane, it was another easy step to get into water skis, and soon DESCO was the leader in this allied field.

The company struggled along through the postwar years under the management of Alfred Dorst. They were not getting rich, but with the sport equipment were successful in holding their own. The name of DESCO was getting to be widely known in the booming underwater sports craze.

Then came Korea.

Suddenly, we were at war again. Uncle Sam was again rearming. The vast war equipment of World War II had largely been destroyed, dumped into the ocean, or sold as surplus. In the new atomic age, new types of warfare were emerging; new types of equipment would be needed.

Dorst was a fine manager and things were getting along well at DESCO. However, with Jack Browne long since gone, having sold the company with Norman Kuehn in 1947, there was no one there who knew anything about diving. The company had continued to manufacture from the same molds, patterns, and jibs the same product that had been made during the war and immediately after the war before Browne left. The resultant equipment was the finest available—but nobody at DESCO really quite understood how it worked or the very important reasons behind the many important details of design.

Dorst asked me one day if I wouldn't like to come back to my old company. New experimental contracts were now in from the Navy Department. A flood of bids was coming in from the government purchasing offices. The surplus had now all but completely disappeared off the market, and commercial business was fine. Water sports equipment was booming, but they would need many new items.

It was a flattering offer, but my heart was still in the deep sea.

One day, I was hanging in the water during a long decompression period and was just thinking. I was cold, tired, and miserable and was wondering just what I was doing in this world so foreign to air-breathing man.

I frankly could not answer the question. The best answer that I could think of was that I still was looking for that elusive pot of gold at the end of the rainbow. It had constantly eluded me. I had always, in the quarter of a century that I had been in the diving business, found enough for a good living and enough to adequately finance my expeditions—but I couldn't ever seem to grasp that fantastic treasure.

I wondered further. What would I actually do if I did find that treasure?

I could not answer that question, either. As I reviewed all of my dreams, hopes, and ambitions, I couldn't think of a single thing that money could buy that I wanted and didn't already have. I tried to visualize that pot of gold right at my feet. I couldn't quite imagine what I would really do with it.

Suddenly, I clearly realized that I already had found that pot of gold. I had a very wonderful and exciting life. Although I had been through countless mental and physical hells, I had spent my entire adult life in a world that I really loved—even though it was a world of which I was still deeply intrinsically afraid. The pot of gold which I had thought that I wanted, and which I now suddenly knew that I had found, was not the actual material gold itself but the essence and substance of the search for it.

I rejoined DESCO. Now, I spend a good part of every day keeping the heavily padded cushion of my overstuffed desk chair warm. An alarm clock wakes me at the same time every morning. Thereafter, the day is on quite a regular schedule.

I am amazed. I had always dreaded this sort of thing. It isn't a bad sort of life after all.

I probably will never wrest that staggering treasure from the sea. I am now convinced that I really don't want it too much. But, as soon as we finish a certain new diving apparatus, now being built in the DESCO experimental department, I know of a little wreck, carrying quite a prize, lying down in the warm tropical waters of the West Indies. She's resting on an even keel just to the leeward of a small palm-fringed island. The sounding on her main deck just happens to be: "by the deep, six."

THE END

Epilogue

The Waters Are Still

By Ted J. Rakstis

M ax Nohl, the man who spent his life chasing after the fortune that he always felt was just within his grasp, never found that pot of gold. Rather, he discovered that gold was not what he wanted.

He and his wife, Eleanor, were on their way back from Mazatlán, Mexico, on February 6, 1960, eagerly looking forward to the start of the *Willem* raising. At seven o'clock in the morning, they passed through Hope, Arkansas.

Max and Eleanor Nohl.

Suddenly, a car loomed up ahead—speeding badly and driving on the wrong side of the road. The end was so quick that Max and Eleanor may have never known it came, and Max probably did not have time to ponder his "I'm out of air, and I don't care" philosophy toward life and death. The Nohls and three of the four passengers in the other car (including singer and pianist Jesse Belvin headed for a show in Dallas) all were killed. Max and Eleanor died when their car burst into flames.

Nohl's death was a shock to many of the citizens of Milwaukee, where he was a legend. The February 6, 1960, *Milwaukee Journal* told the story of the tragic accident and, despite Nohl's death, plans for the raising of the *Willem* went on.

At 3494 North Downer Avenue, Milwaukee, just a few blocks away from Nohl's beloved Lake Michigan shoreline, the friendly voices of Max and Eleanor and the persistent bark of their little dog, Penny (they had no children), no longer are heard. But the books and letters in Nohl's den, the countless pictures and statues of diving scenes, the compasses and ships' bells on the wall—all these are silent witnesses to the life of devotion that Nohl gave to his profession.

And his works live on today, including the self-contained diving suit, now widely used by the US Navy, and the popular skin-diving equipment that might never have been invented had Nohl not worked out the principle to make it possible. In Hollywood, underwater movies are filmed through his cameras. And while his diving record of 1937 now looks small, it was Nohl who made others aware that the water world below could be conquered, to the benefit of all humankind.

These are Nohl's memorials—and they may stand until the seven seas he set out to master have been dried up and forgotten.

Afterword

This book was transcribed from a manuscript submitted to the literary agency Franz J. Horch Associates in New York by Max Gene Nohl and his collaborator Ted J. Rakstis in January 1960. The typewritten manuscript ended, as this book does, with Book VII: The SS *Tarpon*. However, Nohl planned to write another set of six chapters, titled "My Ship Comes In," to end his memoir. These chapters, which Nohl summarized in an expanded table of contents, covered the events of his life between 1953 and 1960, a period as full of excitement as the earlier chapters of his life. Because Nohl was never able to write those chapters, this afterword—written by volunteers and staff members of the Wisconsin Historical Society and supplemented by quotes from Nohl's chapter summaries—provides a brief description of the events that took place during that time.

In 1954, Max Nohl founded the sport diving equipment supply business American Diving Equipment Company (ADE) in Milwaukee. According to an announcement in the June 1954 issue of *Skin Diver* magazine, the company claimed to offer "the most elaborate and complete line of sport diving equipment available in the world today." Nohl remained the technical advisor of Diving Equipment & Supply Company (DESCO), the company he formed with Jack Browne in 1937, and he distributed DESCO products through ADE. To promote the new business, Nohl partnered with George Gross, the proprietor of the Aqua Shop in Milwaukee, to build a four-thousand-gallon plexiglass exhibition tank for the Wisconsin State Fair in the summer of 1954. At the fair, members of the Milwaukee Aqua Club displayed skin-diving methods and gear and even reenacted the underwater picnic scenes from the Grantland Rice Sportlight movie *Shooting Mermaids*, one of the first underwater films, which had been shot by Nohl in 1941.

"I was settling down to a quiet, sensible businessman's life," Nohl wrote. "Then a television announcement changed all my plans." It was

This image accompanied the short article "Aqua Tank Used in Wisconsin State Fair," which ran in *Skin Diver* magazine's November 1954 issue.

an announcement of the sinking of the *Prins Willem V*. At 5 p.m. on October 14, 1954, three miles outside of Milwaukee Harbor, the Dutch-flagged ship *Prins Willem V* was outbound when it collided with the towing cables of the fully loaded Sinclair Oil Refinery Barge 12 as it was being towed by the tug *Chicago*. All thirty men aboard *Prins Willem V* were removed from the ship by the US Coast Guard cutter *Hollyhock* while their ship sank in eighty feet of water.

Nohl lived just over five miles as the crow flies from where the vessel sank. The US Army Corps of Engineers determined that part of the wreckage was a hazard to navigation. "When I learned that the US Army Corps of Engineers was taking bids on salvage, I submitted a bid as a long-shot gamble," Nohl wrote.

Nohl formed the Prins Willem Salvage Syndicate to obtain the financing required to bid on the salvage contract. The syndicate was made up of six Milwaukee businessmen whose names are not known. The contract they entered into with Nohl specified that they would furnish financial backing to obtain a $440,000 bond needed to bid on the job, $30,000 cash to be used as working capital to perform the salvage work, and advice and counsel to assure completion of the contract. Nohl's goal from the beginning was not to remove shallow features of the wreck, but to refloat the entire ship and its cargo, patch the gaping hole, and tow it to Milwaukee where it could be repaired and then sold for profit. Nohl stated that he had no intentions of making a profit from the original bid. Rather, his intent was to submit a low bid to obtain the contract, thus allowing him to gain ownership of the *Prins Willem V*. Nohl won the bid at $50,000 and gained the rights to ownership of the wreck.

Nohl's focus shifted away from ADE after he won the contract. "I decided to put ADE in mothballs," he wrote. Unfortunately, a fire took place at the ADE offices on November 19, 1955. In a letter from John Craig to James Lockwood written nearly thirty years later, on January 25, 1985, Craig wrote that Nohl told him the suit used for his record-breaking 420-foot dive in 1937 was destroyed in the fire.

From the start of the 1955 season, Nohl had his crew prepare the *Prins Willem V* for lifting. According to Nohl, the "Corps of Engineers's soundings [said the ship was] only thirty-one feet below the surface, but I began to suspect it was deeper." Nohl planned to remove the ship's four wooden hatch covers and replace them with metal covers, which could withstand the extreme pressure of the air he planned to pump into the hold to lift the vessel. He also had ventilator covers constructed and installed to make the ship's hull airtight. Another problem was the gangplank, which had loosened from its fastening points, with its long end now hanging down in front of a hatch. It would have to be removed to allow access to the hatch cover.

Midway through preparations, however, the syndicate implemented its right to stop salvage work when funds ran low. They changed the plan and hired the nationally known salvager and demolition expert John Tooker. After surveying the site, Tooker estimated it would take six thousand dollars plus expenses to demolish the wreck. Nohl was not happy and protested. He wanted the ship to remain intact.

In July 1955, the Army Corps of Engineers made another sweep of the wreck to check *Prins Willem V*'s depth. They discovered the wreck had settled fifteen feet since their initial soundings. They surmised that an offshore current sweeping against the wreck and natural settling had done what was desired in the first place. Others suggested there were errors in the corps's initial soundings. Then, Nohl wrote, "we took accurate soundings and found the ship was forty-one feet, six inches below the surface."

Nohl removed the ship's gangplank in the spring of 1956, claiming it was the sole piece of the wreck that was a hazard to navigation. The work turned out to be incredibly easy and fulfilled the stipulations of the contract. The corps felt that the work didn't warrant the pay. Even though the wreck was now deep enough for navigation, the corps argued that since all of Nohl's work was in preparation for raising the ship and not removing the parts that obstructed navigation, he had not fulfilled the contract.

As Nohl described the sequence of events, "The press berated the Army for its costly mistake in judgment, and both the Army and the Syndicate refused to give me my due. A year and a half passed, with most of my time spent in lawyers' offices. Finally, the Syndicate stepped out of the picture. I settled out of court with the Army for $47,000, and I had clear title to the two-million-dollar *Prins Willem* and her cargo."

In 1956, after winning the battle with the corps, Nohl organized a marine salvage company called P-W Corporation (standing for Prins Willem) with Carl A. Backlin of Burlington, Wisconsin. Backlin provided the funds for salvaging the *Prins Willem V*, including supplies and materials.

The first attempt to raise the ship was made in 1956. After months of planning, Nohl converted a fifty-foot cabin cruiser, *Mischief II*, into a salvage tug. Also employed were the tugs *Vitalis*, *Billie*, and *Drifter*. Nohl and Backlin hired an eight-man crew including a teenage diver named Scott Kuesel. Other local salvage divers included Wayne Peltolla and Orville Halversen. The shore station was located at the Allen Bradley parking lot on the east side of Greenfield Avenue.

Nohl changed his strategy for raising the ship from chaining the hatch covers and inflating the wreck to placing surplus bridge floatation buoyancy bags inside the wreck and inflating them. Fifty government-surplus rubber bridge pontoons were to be placed inside the *Prins Willem V*'s four cargo holds to lift the freighter by pumping air into the pontoons and any ship cavities that would hold it. Despite their efforts, the *Prins Willem V* remained on the bottom of Lake Michigan at the end of 1956, mainly because they could not overcome the buoyancy of the pontoons to get them down to the wreck.

In April 1957, Nohl and Backlin started N-B Sales Company, a partnership for the management of employees and day-to-day operations involved in raising the ship. The N-B (standing for Nohl-Backlin) Company was to provide the funds, equipment, and independent contractors for the work. The independent contractors provided the necessary labor for diving needed to salvage the cargo of the *Prins Willem V*. They did not receive compensation for salvaging the wreck itself but shared in the proceeds of the sale of the salvaged cargo, which included outboard motors, machine tool parts, aluminum kitchen ware, and a printing press, among many other items.

In May 1958, Nohl and Armando Conti formed Seaboard Excavators, Inc., for the sole purpose of salvaging the *Prins Willem V*. Subsequently, the P-W Corporation and N-B Sales Company fizzled out. Conti was president of AAA Salvage of Trenton, New Jersey, and he owned the rights to the wreck of the luxury liner *Andrea Doria*, which sank in 290 feet of water off New York in 1956. He traded a portion of his rights to the *Andrea Doria* with Nohl for a portion of the *Prins Willem V*. Seaboard Excavators hired Richard A. Meyers, a renowned salvor and president of Marque Marine Company of Wyandotte, Michigan, to head the *Prins Willem V* salvage operation. Meyers viewed the *Prins Willem V* project as a prelude to the raising of the *Andrea Doria*. Nohl stayed on only as a technical advisor.

Months into this salvage operation, for which ten thousand dollars' worth of equipment had been purchased, Meyers ran off with a barge being used for the project to begin work on a contract to raise the oil-laden tanker *Cleveco*, which had sunk in Lake Erie. This resulted in the US Marshals chasing down the barge on Lake Michigan. As Nohl described it, "Meyers committed technical 'piracy on the high seas' when he absconded with our salvage barge and all my equipment. A federal marshal and I pursued the fleeing barge at night, lost it when Meyers changed course, and finally overtook it forty miles from Milwaukee." The barge was forced ashore in the Sheboygan River, eventually resulting in threats of a lawsuit from the City of Sheboygan unless Seaboard Excavators removed the partially sunken and decrepit barge.

On November 1, 1958, the *Milwaukee Journal* ran an article regarding the shutdown of Seaboard Excavators. After spending two hundred thousand dollars with no results over the summer and fall of that year, the corporation went bankrupt. On December 8, 1958, Nohl resigned from Seaboard Excavators. To recover his dive gear, welding equipment, air compressor, and lost wages, Nohl was compelled to sue the company. A forced auction took place.

In 1959, no known effort was made to raise the *Prins Willem V*. But on October 22 of that year, the *Milwaukee Journal* reported that Nohl had formed a new group called Willem Salvage Corp. for a 1960 attempt at salvage. Shares in the new company were divided among Max Nohl, Robert Meissner of Meissner Engineering Company of Chicago, and Herman Wilms, also from Chicago. As with Seaboard Excavators, the corporation was formed for the sole purpose of salvaging the *Prins Willem V.* "I leave for a winter vacation in Mexico, eagerly awaiting the spring of 1960," Nohl wrote at the end of this chapter summary. "I feel that rainbow's end, and the pot of gold I have been seeking my whole life, is just a few months away."

Max and his wife, Eleanor Nohl, left for their vacation in Mazatlán, Mexico, shortly before Thanksgiving. Nohl planned to rest on the trip before renewing efforts to raise the *Prins Willem V*. Eleanor flew back to Milwaukee for a two-week period for her editorial job but rejoined Max before Christmas for the remainder of their vacation.

A final note on the chapter summaries document, written by Ted J. Rakstis, states:

> It is a virtual certainty that Nohl will raise the *Prins Willem* in the spring of 1960. Two major television networks now are bidding for rights on a "spectacular" on its raising. The story of how it was raised, as the final and crowning touch to Nohl's career, will cover the last chapter. It will point out that even with the ship raised, Nohl's fortune doesn't really matter. Even if he had not raised the ship, he knows that he has already found "the pot of gold at the end of the rainbow"—the gold he sought has been the substance and essence of the search for it. If he somehow should fail to raise the ship, only the last chapter need be changed to stress this point: even though his "moment of glory" eluded him, he has found his life of adventure the only existence he needs for a completeness of self.

Nohl and Rakstis submitted the manuscript and supplementary materials to Franz J. Horch Associates on January 18, 1960. At 7 a.m. on

Nohl poses with a model of the *Prins Willem V* in March 1959.

February 6, while returning from Mexico, Max and Eleanor Nohl were killed in a head-on car crash. The accident happened a few miles outside Hope, Arkansas, on US Highway 67. A passerby managed to pull Eleanor from the burning 1950 Buick, but she died shortly afterward. Several others attempted to remove Max but failed. The family dog, Penny, survived the crash but died the next day.

In the other vehicle, a 1959 Cadillac, was rhythm and blues singer Jesse Belvin, who died at the scene along with his friend and driver, Charles Ford. Belvin's wife and manager, JoAnn, died later at the hospital. It was reported that another performer, Kirk Davis, was in Belvin's car, but it is not clear if he survived. The state troopers reported that two tires had been slashed on Belvin's car and that he had received death threats from the Ku Klux Klan while performing in Little Rock four hours before the accident. At the time of the crash, Belvin's new black Cadillac swerved into the oncoming lane. The official cause of the accident stated that Ford, the driver of the Cadillac, fell asleep at the wheel. Previously, Ford had been fired from another group for falling asleep while driving.

The *Prins Willem V* has never been raised, though several attempts were made in the years following Nohl's death. Today, it is without question the most visited shipwreck in Wisconsin waters. The wreck is close to Milwaukee Harbor, and a plentitude of dive charters operate from the city. With typically clear water conditions and a reasonably shallow depth, it often makes scuba magazine lists of the top five dive sites in the United States. Since the passing of the Abandoned Shipwreck Act of 1987, this well-known wreck diving destination has been owned by the State of Wisconsin and managed by the Wisconsin Historical Society for all to enjoy.

The Special Collections of the Milwaukee Public Library

The Max G. Nohl Papers are part of Milwaukee Public Library's (MPL) Local History Manuscript Collections (LHMC), one of several of MPL's distinct Special Collections. The LHMC's focus is on personal and family papers, as well as records from Milwaukee-area organizations, businesses, and significant local events. It was created in the 1950s, when MPL started its Local History Room and solicited manuscripts from local authors. After Nohl's death in 1960, his papers were donated to MPL by J. Gordon Hecker, the brother of Nohl's wife, Eleanor (Hecker) Nohl. They were later processed by MPL staff and included in the LHMC.

Milwaukee Public Library's Special Collections contain a variety of materials and cover a wide range of topics, such as politics, genealogy, sports, music, and more. A sample of the individuals and organizations included are Frank Zeidler, Irene Goggans, Edgar End, Erich & Lucia Stern, Billy Mitchell, Elizabeth Corbett, August Derleth, the Wisconsin Visual Artists, the Milwaukee Turners, the Socialist Party, and the Milwaukee County Republican Party. Collections are used by researchers, students, authors, filmmakers, local history enthusiasts, media outlets, and others. In addition to the LHMC, MPL's Special Collections include the Wisconsin Architecture Archive, Milwaukee Road Archive, City of Milwaukee Archives, Historic Photo Archives, and Great Lakes Marine Collection, which is supported by the Wisconsin Marine Historical Society. Finding aids for the LHMC and other Special Collections can be found online at Archival Resources of Wisconsin.

To learn more, visit:
https://mpl.org/archives/local_history_manuscript_collections.php.

INDEX

Page numbers in **bold print** refer to illustrations.

A. Schrader's Son, 243

AAA Salvage, 379

Abandoned Shipwreck Act of 1987, 382

agglutination, 191, 254, 263

Alligator Boat, 274, **274**

American Diving Equipment Company, 375, 377

Andrea Doria (liner), 379

Andrew J. Morse & Son, Inc., 28, 242–243

Antietam (Coast Guard cutter), 206, **207**, **210**, 220

Apalachicola, FL, 307, 317

Apalachicola Steamship Company, 333, 365

Aqua Shop, 375

Aquaplane, 370

argon, 196

Arna (diving boat), 337, 346, 353

Australia (cruiser) 93–94

Backlin, Carl A., 378–379

Backman, Carl, 239

Baker, Adley, 335

Ball, Edward, 267

barracuda, 310–315

Barrow, Willis Green, 333–334, 365–366

Barry, Norman, 223–225, **224**, 228–229, 232–235

Barton, Otis, 57

Bathysphere, 55–56, 57

Bechstein's Swimming School, 15

Beebe, William, 55–56

Belvin, Jesse, 374, 381

Belvin, JoAnn, 381

bends, the, 157, 178, 196, 198, 211, 252–254, 292–293, 348–356

agglutination, 191, 254, 263

research, 191–203, 254–264

Billie (tug), 378

British Admiralty Supersonic Echo Sounder, 213

Brown, Prescott, 62–67, 70

Browne, Barbara, 244–246

Browne, Jack, 17–18, **20**, 98, 186, 243–246, 327–330, **328**, **329**

and DESCO, 243, 246–247, 327, 368, 370–371, 374

and diving record, 327–330, **328**, **329**

and Lusitania Expedition, 185–186, 212, 246

and self-contained diving suit, 189, **190**, **193**, 204–206, 208, 249

buoyancy, 55, 139–140, 155, 156, 226, 271, 272, 292, 360–361, 378

negative, 56–57, 192

positive, 56

zero, 56–57, 298

Buzier, Kosta, 307–309

cameras. *See* underwater cameras

Captain Pete, 286–290, **289**, 293–295

carbon dioxide, 253–254

Carmichael (Captain), 100–103, 152

Chase, Don, **235**

Cherry (crewmate), 336, 343, 346–349, 353–356, 364, 366

Chicago (tug), 376

cinematography. *See* motion pictures; photography; underwater cameras

Cleveco (tanker), 379

Columbia (steamship), 89–92

communications, 175, 279, 301–305, 354–355
    diver/ship, 138–139, 142, 144–147, 154, 156–157, 163–164, 170, 171–172, 231, 340–341
    radio broadcasts, 77–78, 79, 83, 206–210, 220, 223–229, 234–235
    shortwave broadcasts, 89, 206, 211, 220

Conti, Armando, 379

Craig, John D., 175–178, **176**, **177**, **184**, **187**, 199–203, **202**, **203**, 377
    and Lusitania Expedition, 183–185, 188, 212, 237, 238
    radio broadcasts, 204, 206–211, **210**

Crilley, Frank, 53, 56, 177, 184, 198, 221, 224, 228, 232, 330

Curney, Dave, 97, 99–100, 105–106, **106**, 109–111, **109**, 113, 117–118, 120, 122–126, 129, **130**

Curney, Donald, **130**

Curney, Hat, 122–123

Cuttyhunk Island, MA, 101, 119, 120–122, 127, 129–131, 134, 149, 152, 161, 174, 175, 177

Danford, L. E., 333–334

Davis, Elsie, 275

Davis, Kirk, 381

decompression
    after dives, 87, 142, 147–148, 157–158, 174, 231, 235, 278, 292, 315, 342, 346, 350–351, 356, 363, 371
    experiments, 198–203, 254–264, 329
    graph, **211**
    tables, xi, 235, 252
    theory of, 196, 197
    tunnel workers, 252–253

Demetriades, H. J., 184, 185

DESCO. *See* Diving Equipment & Supply Company

Dexter, KS, 197

diving, **v**, 367–368. *See also* dives *under* Nohl, Max Gene
    mixed-gas, x, xi, 178, 186, 191, 196, 199–200
    rate of pay, 49–50, 125

diving accidents, 45–48, 198, 208–211, **210**
    and *Antietam* (Coast Guard cutter), 225, 229–230
    and *Empire Mica* (tanker), 314–316
    and *John Dwight* (steamship), 143–148, 153–154, 156–157, 171–174
    and *Tarpon* (steamship), 341–346, 358–362

diving equipment, ix–x, 89, 94, 185, 242–243, 251, 367–368
    bells, 15, 55–57, **58**, 65, 79, **85**, 86, 87–88, **87**, 94, 279, **279**
    helmets, **4**, 17–25, **17**, **20**, 26–27, 28, **28**, 34–36, **37**, 42, **43**, 62, **137**, 140, 144, 185, **187**, **195**, 207, 311, **369**
    ladders, 21–22, 136, 188, 279, 311, 315–317
    lung, xi, 298–299, **299**, **300**, 302–303

Diving Equipment & Supply
  Company (DESCO), ix-x, xi, 247–
  249, 327, 329, 367–372, **369**, 375
diving records
  Browne, 327–330, **328**, **329**
  Crilley, 53, 56, 177, 184, 198, 221,
    224, 228, 232, 330
  Nohl, x, xiii, 219–236, 246, 263,
    327, 328, 330, 374, 377
diving suits, xi, **4**, 20, 26–27, 28–29,
  28, 33–34, 36–37, **37**, 42, **43**,
  44–48, 54, **137**, 153, 205–206,
  205, 243, **245**, 288, **289**
  Lightweight Suit, 327–329, **328**
  self-contained, 56, 97, 178, **179**,
    180, 183, 185–206, **187**, **190**,
    193, **194**, **207**, 212, 219, **222**,
    224, **226**, 237, 242–248, **245**,
    263, 284, 299, 374, 377
  steel, 54–55, **55**
diving tower, 272, **273**
"Doc" *See* Hoffman, Norman "Doc"
Dorchester (freighter), 102
Dorst, Alfred, 368, 370–371
Drifter (tug), 378
DuPont, Alfred I., 267
Dussaq, René, 175–177

echo sounder, 213
Empire Mica (tanker), 306–318
End, Edgar M., x, xi, xiv, 191–192,
  199–203, **202**, **203**, 212, 219,
  223–224, **224**, 231, 235, **236**,
  254–262, **257**, **262**, 329, 383
End, Kathy, xiv
Erwin, Russ, 275
exhibition tank, 375, **376**

Fischer, Joseph, 198–200, **201**, 255–
  256, 259–262

Flink, Constantin, 66–68, 72, 74,
  79, 80
Ford, Charles, 381
Fox Point, WI, 19, 219
Franz J. Horch Associates, xiv, 375,
  380
Fry, Ken, 204, 220, 223

gasoline
  rationing, 319
  salvage, 307, 318
Gay, Maurice, **365**
Glass House, 272–274, **274**
glass-bottom boat, 268, **268**
Grantland Rice Sportlight, 274–276,
  375
Greenough, Vose, 29–30, 32–36, 38,
  41, 94, 97–98, 151–156
Gross, George, 375
guinea pigs, 192, **192**, 198–199, 224,
  263

Halversen, Orville, 378
harpoon gun, 300, **301**
Hecker, Eleanor. *See* Nohl, Eleanor
  Hecker
Hecker, J. Gordon, xiii, 383
helium, 221, 224–225, 231, 236
  experiments, 191–204, 224, 254–
    264, 327–330
  and self-contained diving suit, 205,
    211, 212, 221
Hermosa (diving boat) 310, **310**, 315,
  321, 324
high-pressure experimental animal
  chamber, 192, **192**
Hoffman, Norman "Doc," 104–105,
  111, 114–117, 126, 130, 132, 148–
  150, 155, 157–161, 166
Hole in the Water, 268, **269**, 277

Hollyhock (Coast Guard cutter), 376
hurricanes, 316–317, 333
hydrogen, 197
hyperbaric chambers. *See* pressure
  chambers

inert gases, 196–197, 198, 254
  argon, 196
  helium, 191–204, 205, 211, 212,
    221, 224–225, 231, 236, 254–
    264, 327–330
  krypton, 197
  neon, 197
  nitrogen, 197–199, 203, 254–255,
    261, 278, 342
  xenon, 196
inventions
  Alligator Boat, 274, **274**
  Aquaplane, 370
  diving bell, 55–57, **58**, 65, **85**, 86,
    87–88, **87**, **279**
  diving ladder, 311
  diving lung, xi, 298–299, **299**,
    300, 302–303
  diving tower, 272, **273**
  Glass House, 272–274, **274**
  Hole in the Water, 268, **269**, 277
  Lightweight Suit, 327–329, **328**
  patents, 185, 191, 247
  self-contained diving suit, 56, 97,
    178, **179**, **180**, 183, 185–206, **187**,
    190, **193**, **194**, **207**, 212, 219,
    222, 224, **226**, 237, 242–248,
    245, 263, 284, 299, 374, 377
  submarine pneumatic harpoon
    gun, 300, **301**
  underwater houses, 268, 270–271,
    270
  underwater parachute, 271–272,
    271

jinx boats, 253–254
JM Allmendinger (steam barge), 98
John Dwight (steamship), 97, 99–103,
  200, 340
John Dwight Expedition, 104–175,
  177, 183
  salvage, 163–165, **165**, 170, 183
  map, **119**, **121**
Johnstone, Ernest. *See* Pōhaku
  (Ernest Johnstone)

Kenney, Edmund, 307
King (Captain), 100–103, 152
King, John, 103
Kopmeier, Barbara, 244, 246
Korean War, 370
krypton, 197
Kuehn, Norman L., 246–249, 368,
  370
Kuehn Rubber Company, 246, 248
Kuesel, Scott, 378

Lake Michigan, 19, 57, 185, 198, 204,
  212, 374
  diving record, x, 219–236, 327
  JM Allmendinger (steam barge), 98
  Norlond (steamship), 206–211
  Prins Willem V (freighter), 374,
    376–380, 382
  Westmoreland (steamship), 186
Larkin, Tommy, 177
lawsuits, 378, 379, 380
Lindsey (crewmate), 309, 311
Lockwood, Jimmy, 281, 377
Lord, Phillips H., 61, 64, 66–72, 75,
  76, 77–85, 89–93
Love, Jack, 89
Lowry, Bunny, 275
Lusitania (liner), 52, 54, 184, 224,
  249–251

Lusitania Expedition, 53, 58, 177, 184–185, 188–189, 212–216, 219, 237, 244, 246, 249, 251

Marque Marine Company, 379
Marquette University, 191–192, 224
Marsh, Truman, 104, **106**, 107–110, 113–118, **130**, 132–133, 135–149, 152, 155–166
Massachusetts Institute of Technology, 27, 41, 52–53, 67, 71–72, 74, 94, 97–99, 104
Meissner, Robert, 380
Meissner Engineering Company, 380
Mendota (Coast Guard cutter), xii
Metro-Goldwyn-Mayer, 276
Meyers, Richard A., 379
Milwaukee, WI, x, xiii, 104, 198, 374, 376, 382
Milwaukee Aqua Club, 375
Milwaukee Athletic Club, 193–194
Milwaukee County Hospital, 198–199
Milwaukee Journal (newspaper), 93, 189, 246, 374, 379–380
Milwaukee Public Library, x, xi, xiii, 383
Milwaukee River, 205, 221
Mischief (tug), 378
Mischief II (cabin cruiser), 378
Moody, Dick, 13–14
Moose Lake, 9–15, 32
Morgan City, LA, 320–322, 324
motion pictures, 88, 92, 175, 177, 237, 247, 268–280, **275**, **277**, 301, 374, 375
novelties, 268–274, **268**, **269**, **270**, **271**, **273**, **274**

N-B Sales Company, 379
National Broadcasting Company (NBC), 68, 72, 75, 78, 204, 206–211, 220, 221, 229
neon, 197
Netzow, Verne, **20**, **98**
news media, 189, 203, 220, 221, 256. *See also* National Broadcasting Company (NBC)
nitrogen, 197–199, 203, 254–255, 261, 278, 342
Nohl, Eleanor Hecker, xiii, 322–326, 373, **373**, 380, 383
death, 374, 381
and diving, 324–326
Nohl, Max Gene, **xii**, **87**, **106**, **190**, **240**, **251**, **310**, **373**, **381**
college, 26–27, 41, 58, 67, 71, 72, 94, 98, 104
death, xiii, 373–374, 380–381
and DESCO, ix-x, 246–249, 367–372
dives, 21–25, 31–41, 50–51, 86–87, **98**, **109**, 128, **130**, 136–149, **137**, 155–158, 160–166, **165**, 169–174, **177**, **187**, **193**, **194**, **205**, 206–211, **207**, 212, **222**, 225–236, **226**, **235**, **236**, 277–280, **279**, 289–293, **289**, **299**, **300**, **301**, 303–304, 311–316, 318, 336–346, 356–364
diving record, x, xiii, 219–236, 246, 263, 327, 328, 330, 374, 377
fear, 10–11, 14, 15–16, 19, 24, 35–36, 42, 65, 105, 111, 171, 233, 237, 260, 303, 317, 338, 348–349, 358, 361, 371
finances, 28, 72, 94, 126, 188, 237, 238, 247, 280, 281, 298, 368, 371, 376–377

lecturing, 237–241, 247, 281, 311, 322, 324

marriage, 322–324

memoir, x, xi, xiii-xv, 375, 380, 383

partnership with John D. Craig, 185–189, 199, 212, 237, 238

research on bends, 191–203, **202**, **203**, 254–264, **257**, **262**

residence, 15, 19, 183, 219, 374, 376

Nohl-Browne Diving Gear, 243–249, **248**

Norlond (freighter), 206–210, 220

Orphir (salvage vessel), 213–214, 216, 219, 244

oxygen intoxication, 197–198

P-W Corporation, 378, 379

Panama City, FL, 306, 333, 335, 353

Parker, Seth. *See* Lord, Phillips H.

Parks, John "Whiskers," 73–74, 76, 79, 80, 92

patents, 185, 191, 247

Peltolla, Wayne, 378

Perry, Newt, **269**, **271**, **274**, 275

photographers, 189, 203, 220, 256

photography, 88, 237–238, 267–268. *See also* motion pictures; underwater cameras

Pōhaku (Ernest Johnstone), 52–53, 117

porpoises, 300–304

Port Washington, WI, 220

pressure chambers, x, 192, **192**, 198–203, **201**, **202**, 255–262, **257**, **262**, 327–330, **329**

Prins Willem Salvage Syndicate, 376–378

Prins Willem V (freighter), 374

model, **381**

salvage efforts, 374, 376–380, 382

Prohibition, 100–101, 167

radio broadcasts, 77–78, 79, 83, 206–210, 220, 223–229, 234–235

Rakstis, Ted J., 375, 380

recompression, 196, 198–203, 253

Redpath, James, 239

Redpath Bureau, 239, 240

reporters, 189, 203, 220, 221, 256

Russell, Henry, 213–216, 219

St. Phillip (sponge boat), 285–287, 289

salvage, 15, 88–89, 97–99, 183, 214, 244, 249, 298, 318

beer, 339–340, 346–347, 362

cash, 183, 365–366, **365**

gasoline, 307, 318

gold, 186

Prins Willem V, 374, 376–380, 382

whiskey, 104–175, 186

white lead, 206

Sea Killers (motion picture), 175

Seaboard Excavators, Inc., 379–380

Search Board, 370

self-contained diving suit. *See under* diving suits

Seth Parker (schooner), 63–65, **63**, 73, **76**, 78, 80, **85**, 86, **87**, 93–94

Seth Parker Expedition, 66–94

sharks, 3–5, **4**, 53, 74, 303–305, 310–311

Sheboygan, WI, 379

Sheffield, Johnny, 276, **277**

shipwrecks, 88–89, 214–215, 281–282, 297–298, 318, 320–321, 326, 382

Abandoned Shipwreck Act of 1987, 382

Andrea Doria (liner), 379

Cleveco (tanker), 379

Empire Mica (tanker), 306, 307–318

JM Allmendinger (steam barge), 98

John Dwight (steamship), 97, 99–103, 200, 340

Lusitania (liner), 52, 54, 184, 224, 249–251

maps, **119, 121**

Norlond (freighter), 206–210, 220

Prins Willem V (freighter), 374, 376–380, 382

Tarpon (steamship), 333–335

Westmoreland (steamship), 186

Shooting Mermaids (motion picture), 274–276, **275,** 375

shortwave broadcasts, 89, 206, 211, 220

Siebe, Augustus, 26–27, 54

Silver Heels (salvage vessel), 98, 99, **106,** 111, 114, 119, 122, 132, 134, **137,** 149, 169, **177,** 183

engine failure, 158–160

Smith, Eugene, 255

Snow, Roland, 101–102, 120, 121–122, 155

sound waves, 213

South Milwaukee, WI, 206

sponge diving, 281–296, **283, 289,** 347

sponges, 290–291

grass, 291

wire, 291

wool, 291

yellow, 291

Stein, Harold "Uncle Hal," 104, **106,** 107–109, 111, 115, 117, 122, **130,** 135, 152, 157–160, 174

storms, 310, 316–317, 334–335. *See also* hurricanes

submarine pneumatic harpoon gun, 300, **301**

Swindle, Percy, 191

Tarpon (steamship), 333–335

Tarpon Expedition, 333, 335–366

Tarpon Springs, FL, 253, 281–285

Tarzan's Secret Treasure (motion picture), 276–277, **277**

Thomsen, Tamara, ix-x, xiv

Tooker, John, 377

Traynham, Floyd, 277, 278, 280

Turner (Captain), 213, 214, 215

"Uncle Hal" *See* Stein, Harold "Uncle Hal"

underwater cameras, xi, 65, 80, 88, 97, **98,** 189, 219, 251, **251,** 298, **299,** 302, 374

underwater houses, 268, 270–271, **270**

underwater parachute, 271–272, **271**

US Army Corps of Engineers, 376–378

US Coast Guard, 99, 101–102, 120, 121, 122, 206, 220, 221, 223, 227, 230

US Navy, xi, 99, 249, 318, 320, 328–329, 368, 370, 371, 374

Vestrem, Ivan, **222,** 223–232, **235, 236, 251**

Vineyard Haven, MA, 99, 122, 126, 151

Vitalis (tug), 378

Wakulla Spring, 267

Walden Pond, 31–41, **37,** 153

Warren, Rod, 275

water sports equipment, 370, 371

Weissmuller, Johnny, 276–277, **277**
Westmoreland (steamship), 186
"Whiskers" *See* Parks, John
  "Whiskers"
Whitman (Captain), 220–232
Willem Salvage Corp., 380
Wilms, Herman, 380

Winnie, Russ, 205–210, **210**
World War I, 249–250
World War II, 249, 251, 306, 313,
  318–320, 367–368
WTMJ (radio station), 204–211

xenon, 196